LIVING IN LIGHT OF GENESIS

A FOUNDATION FOR FAITH, SCIENCE, AND CULTURE

LIVING IN LIGHT OF GENESIS

A FOUNDATION FOR FAITH, SCIENCE, AND CULTURE

BRIAN THOMAS, PH.D.

Dallas, Texas
ICR.org

LIVING IN LIGHT OF GENESIS
A FOUNDATION FOR FAITH, SCIENCE, AND CULTURE

by Brian Thomas, Ph.D.

First printing: December 2023

Copyright © 2023 by the Institute for Creation Research. All rights reserved. No portion of this book may be used in any form without written permission of the publisher, with the exception of brief excerpts in articles and reviews. For more information, write to Institute for Creation Research, P. O. Box 59029, Dallas, TX 75229.

All Scripture quotations are from the New King James Version.

ISBN: 978-1-946246-74-5
Library of Congress Catalog Number: 2023950991

Please visit our website for other books and resources: ICR.org

Printed in the United States of America.

Published by Wayfinders Press, an imprint of ICR Publishing Group

For Dad.
He met Jesus at age 80. His last words to me in the hospital were "Don't stay up all night writing your book." I didn't. I wish he could have seen it. Even if history gives it a vote of mediocre, he would have been proud of it just because I'm his son. You can't beat that.

TABLE OF CONTENTS

Acknowledgments 9

Prologue, Sort of 11
    A Garden in Exodus? 11
    A Garden, Its Maker, and Me 13
    Why This Book? 14

1. What Genesis Says 17
    Our Tension with Genesis 17
    What's at Stake with Genesis? 19
    Genesis 1 19
    Who God Is 21
    My Struggles with Genesis 26
    Sin and Death 27
    Judgment and Grace 28

2. Genesis and Evolution 33
    To Contradict or Not to Contradict 33
    Side by Side 34
    Cut and Paste 36
    Free Science from Moses 42

3. Science That Supports Genesis 47
    The Power and Weakness of Science 47
    Astronomy and Genesis 54
    Genetics and Genesis 61
    Biology and Genesis 68
    Geology and Genesis 77
    Fossil Proteins and Genesis 78
    Science and Bias 82

4. Culture Without Genesis 89
    What People on the Street Believe About Genesis 89
    What Christian Intellectuals Believe About Genesis 94

What Bible Believers Say About Genesis 101
Communism's Position 103
Genesis in Germany 104
Genesis and Hot Buttons 107
Genesis and America 120
Culture Minus, or Plus, Genesis 122

5. Genesis and the Bible 123
Five Core Concepts That Lean on Genesis 123
Five Unbroken Threads That Wind from
    Genesis to Revelation 129
What the Prophets Thought of Genesis 141
What the Apostles Thought of Genesis 144
How Jesus Treated Genesis 147

6. Genesis and Truth 149
Where to Draw the Line? 149
Genesis and the Gospel 152
On the Origin of Our Ugly Insides 159
Genesis and Meaning 162

7. Why Genesis Matters More Than Ever 167
The Genesis Disconnect 167
Genesis and Everyday Life 169
Dwindling Christianity 171
The Limp Church 174
Joys from Genesis 176

Endnotes 183

Image Credits 189

About the Author 191

ACKNOWLEDGMENTS

Jayme Durant, my former boss, friend, and Director of Communications at the Institute for Creation Research, and her excellent editors read what I wrote in this book more carefully than I did. If anyone finds this book in any way relatable, it's because of them.

The late Dr. Beatrice Clack at Stephen F. Austin State University, my former supervising professor who was energy and willpower incarnate, believed in 1999 that I could write an entire master's thesis, so I did. I doubt I could have written a book like this one without that milestone.

Professor Stephen Taylor of the University of Liverpool, my other former supervising professor excellent preacher and pastor, mission organization (or to him, organisation) president, business owner, research department head, expert instructor, and a man who sees what can be believed, said in 2015 that I could complete a Ph.D. under him, so I did.

After my graduation, we prayed for the "genesis" of a book like this one. My dear wife, Michele, magnificent mother of five, executive director of a nonprofit, servant of all, and constant companion, sticks with me despite my flaws. Her love for Jesus inspires me to do things like write books about how great God's Word is.

PROLOGUE, SORT OF

A Garden in Exodus?

Sitting in church, listening to our pastor's sermon on Exodus, I was picking up the story arc he was putting down. There's just so much in the Bible—so many connections and ties, nuances and overt truths, stern warnings and gentle wisps of insight or encouragement. And Pastor Matt was showing how another one of these gems ties us back to the Garden of Eden and forward to a future garden, complete with its very Builder live and in Person.

He was preaching from Exodus, yet he called our attention to Genesis. Yes, that is the book most folks think of as at least partly myth. I did too, once. Now I believe its words, and I think the Garden of Eden really existed. So, I wanted to write a book that glimpses into what made me do a 180 on Genesis. A book about how a real garden—along with creation and the Flood—could impact our everyday and everlasting lives.

Genesis says of this ideal place, "The Lord God made every tree grow that is pleasant to the sight and good for food" (Genesis 2:9). Details are sparse, but that just adds intrigue. It had trees that looked gorgeous and offered free, delicious food whenever people felt hungry. Sign me up.

And the Lord Himself was there, "walking in the garden in the cool of the day" (Genesis 3:8). There was total joy in the presence of the Maker—the One who invented joy. His very nature, being three Persons who each share the same essence, defines harmony. But the harmony between God and man didn't last long according to this narrative. Our original ancestors, despite this gorgeous garden and its grand Architect on hand, decided to distrust God. The first husband and wife rebelled. Now each of us inherits that same fist-shaking nature even as we each inwardly hunger for harmony with others, with God, and even within ourselves. What a mess.

Ever since that first sin, what son or daughter of Adam and Eve wouldn't love to find a way out of this mess and back to that good garden? It represents relationships mended, grace that overcomes our rebellion, wrongs made right, wounds healed, and perhaps most precious of all, God's wholehearted acceptance of who we are despite who we are.

Back to that sermon. The scene in Exodus 3 describes God calling Moses to help free His people from slavery in Egypt. Moses offered a string of excuses for how and why he should not be the one to do this. Verse 12 says, "So He said, 'I will certainly be with you. And this shall be a sign to you that I have sent you: When you have brought the people out of Egypt, you shall serve God on this mountain.'" What? How did a mountain make its way into the text? How does it relate to what's going on?

Pastor Matt recalled Genesis 2:10: "Now a river went out of Eden to water the garden, and from there it parted and became four riverheads." He explained that since water flows downhill, God, Adam, and Eve must have shared those first, innocent moments amidst an elevated garden. A God-made mountain. His exalted dwelling place on Earth.

When this same God offered Moses a sign, He hearkened back to the garden—that original mountain where He intended His creations to know and enjoy Him with ever-unfolding delight. The pastor's next slide read, "The story of the Bible is not about making life in the wilderness better for us. It is about something much better—going back to Eden." What makes Eden better? That was and will be, but for now represents, the place where we get to know our Maker.

Pastor Matt noted the great promise of that future mountain. The Bible foretells the form of an immense, golden, God-crafted city. Just as the waters flowed from the center of the garden in the beginning, this new version features a "pure river of water of life, clear as crystal, proceeding from the throne of God and of the Lamb" (Revelation 22:1).

This water will descend from His heights to nourish all those who let the Lord Jesus take care of the sins that separate us from our Maker.

A Garden, Its Maker, and Me

God's promise to Moses about a mountain ties Moses into the story arc of a return to a garden with God in it—available to chat. When we read what God said to Moses, "When you have brought the people out of Egypt, you shall serve God on this mountain," we might hear Him pointing us to Himself as if to say, "When you have completed the labor of your life, you shall serve God on a blissful garden mountain like it was in the beginning."

For that matter, the many Bible references to a future Mount Zion weave new threads into this tapestry. That is where God promises to dwell with His people in harmony once again. Even Isaiah unspools such a thread, saying:

> Is it not yet a very little while
> Till Lebanon shall be turned into a fruitful field
> [i.e., a garden full of produce],
> And the fruitful field be esteemed as a forest?
> In that day the deaf shall hear the words of the book,
> And the eyes of the blind shall see out of obscurity and out of darkness.
> The humble also shall increase their joy in the LORD,
> And the poor among men shall rejoice
> In the Holy One of Israel. (Isaiah 29:17-19)

There He is—the Holy One with us in joy, together in a garden. And that began in Genesis.

Whenever I get the chance to enjoy parks, I try to take it. I've walked all over the country. The ponds, paths, and people elicit praise. Such escapes from my man-made world of cars and concrete bring me back to God's world and ultimately toward Him. But He's not physically beside me in those walks—not yet—even though His presence in spirit is even closer than that. The wonders of creation in such gardens beckon me to honor the One who deserves all the credit—but He remains invisible! It's like a cheering audience after the curtains close on a two-act play, only this time the cast and crew stay backstage! Something's missing on my walks. If God were physically there when I feel the softness of a rose petal, I might be able to see how His face reacts when I exclaim, "Well done and wow!" right to its Maker! Maybe we would watch a big-eyed beetle, and I could ask, "So, what were you thinking when you formed this one?" Setting aside those creations, I could ask, "Why do you care so much about someone so small as me?"

Like the cast and crew returning to the stage to receive their well-earned praise, our great Creator's return to our world—and why not into a garden fit for a king?—promises to fulfill my longing for a true and rightful recipient of my expressions of wonder and admiration. Did this longing find its origin in the pages of Genesis? Could it be that one day you and I will find the ultimate joy in an ultimate relationship with Jesus the Creator as we walk and talk like we were meant to?

The same Creator who handcrafted each of us to know and enjoy Him forever provided a living Way to bring us back to Himself. Of course, we must lower those raised fists to begin a relationship with One so holy. The how, who, and why of that Way to God and His garden of completion and peace begin in Genesis. These questions suggest—much as my pastor did on that Sunday morning—we can return to the God of the garden of old and so partake of His new garden one day. This motion toward the Creator means motion toward His book of beginnings. Genesis not only offers a sketch of where we came from but sets up themes that wind through the rest of God's Word and maybe even through our own lives.

Without Genesis, where can we find a foundation for the basics about walking and talking with God or what went wrong to break that fellowship in the first place?

Why This Book?

Now I hope you see my motivation for writing this book. When I talk about a return to the garden, I do not mean we should go backwards. Quite the opposite. I'm talking instead about plugging into a story more universal, fundamental, and epic than that of our brief lives, and yet one that our very lives can help write!

It is hard to plug into this story when our culture knows less and less of what is true. Through Hosea, God said that "My people are destroyed for lack of knowledge. Because you have rejected knowledge, I also will reject you from being priest for Me; Because you have forgotten the law of your God, I also will forget your children" (Hosea 4:6). We see so many of the consequences of this lack of knowledge today. I have experienced some of them as I'm sure we all have.

I found no easy way to plug into the great story back when I discounted and doubted the Garden of Eden, Adam, Eve, temptation, sin, redemption, grace, and all that Genesis lays out so clearly. For that matter, I knew I had no answers to society's plagues such as why so many marriages fail, tough conundrums like why evil exists, tragic histories like the Holocaust, or why we can feel so empty. Then through the study of stars, rocks, genetics, history, logic, and the Bible itself, I started to plug in. I wish in these pages to make a case for why everyone should take a closer look at the historicity, reliability, and relevance of Genesis as a firm foundation for understanding history, interpreting scientific observations, rebuilding strong families, knowing ourselves, and ultimately knowing God. Living in light of Genesis gives us this foundation.

1
WHAT GENESIS SAYS

"But if you cannot understand how this could have been done in six days, then grant the Holy Spirit the honor of being more learned than you are."[1]
— *Martin Luther, ca. 1530*

Our Tension with Genesis

The book of Genesis begins with the most widely known passage of the entire Bible. "In the beginning God created." Nothing there about natural processes that take long time spans. When I read it as a youngster I was already indoctrinated into nature-only philosophy. These culture-soaking ideas distance us from miracles. No wonder the Genesis assertion of a bona fide divine miracle sounded so strange. Hadn't science buried miracles long ago? Later chapters of this book unpack what most folks mean by *science* as well as what Westerners think about miracles. But for now, suffice it to say that those who approach Genesis, as I once did, with a predisposition against the very possibility of miracles have a hard time swallowing Genesis as history.

Do you see the core of my struggle?

In discussing Genesis over the years, I have heard and considered some wild attempts to resolve this tension. This book is all about how I found resolution. As such, it explains a bit about what fuels my passion for going back to Genesis.

A regular church attender reacted to one of my creation presentations and followed up with a letter. It displays the strain we so often feel when reading Genesis with modern eyes.

> I think that is why you either accept the Bible as literally true and accept that it strongly conflicts with historical, archaeological, and scientific knowledge for reasons we simply can't understand because God didn't reveal why that is the case, or you accept that the Bible isn't literally true in all regards but some parts may be sacred. We just don't know what parts those are, thereby rendering the whole Bible open to doubt that a Divine hand had anything to do with it.[2]

I think he lays out the basic logical options. But if we can't know which parts of the Bible to call sacred, then what use is any of it?

The author of this letter may mistake conventional scientists' confident opinions about the past for real science. That's what the Institute for Creation Research, where I work, is all about—removing sciencey-sounding objections to biblical Christianity and replacing them with the abundant observation-based science that confirms Genesis. By the end of this book, I hope to persuade you to join me in the journey that led me to accept the Bible's history as literally true not *in spite* of historical and scientific knowledge but *in light* of that knowledge.

I remember a conversation I had in my office with a man in his twenties who had recently been born again in Christ (1 Peter 1:23). He expressed his worldview struggles and the same kinds of origins questions that I wrestled with for years. I asked him what he believed about Genesis. His response seemed to match many modern Christians. He repeated the culturally accepted notion that Genesis was not written for the purpose of conveying actual historical events. Instead, he suggested it was to establish the God of Israel as being more worthy of worship than the gods of other nations. Perhaps He is a god who lurks somewhere in the background of the universe. But my friend would say He is not a god who conveyed straightforward history in Genesis 1 since science had supposedly disproven that history.

Was the creation account merely written as a kind of false historical frame to elevate Israelite esteem for God? How worthy of worship would they have considered a god who needed such contrived bootstraps? Would the Israelites have fallen for that kind of shtick? Would you? Are we certain of the creation account author's intent? How do we know for sure?

I want to express the answers I have found to these and other Genesis questions in this book. Back when I first read Genesis, I felt incredulous. I balked at some of its wild-sounding assertions. It took a lot of study for me to begin to find resolution. I invite you to compare your journey with mine. In the words of my dear former supervising professor at the University of Liverpool, "Give it a think."

What's at Stake with Genesis?

What's at stake with knowing the truth about our origins? Who on earth doesn't want to know—at one point or another—where he or she came from? The consequences of getting the wrong answer to origins affect daily decisions. Such decisions even echo in eternity, since we humans have spirits that last beyond these mortal bodies. If we evolved from animals with no divine oversight like most naturalistic scientists teach, then how does that teaching affect our daily moral choices? Do we feel justified in valuing our neighbors as little more than hairless apes or in serving them as highly valued humans?

On the other hand, if God created the first people, then how does that alter our outlook? Do we or should we seek to learn more about this Creator or even go so far as to align ourselves with His expectations? The contrast is stark. The question of origins does not belong to distant and dusty halls of philosophy. It impacts every moment, directly or indirectly. Questions of origins and how we regard Genesis carry vital importance for many people—people like you and me.

Genesis 1

If we are going to do this right—if we seek to rightly judge the degree to which Genesis matters—then an honest look at what Genesis says right on its surface should come first. If we read Genesis with a simple mind—untainted by indoctrination—what would we make of its message? Let's go back and view the beginning through the eyes of a child.

> The earth was without form, and void; and darkness was

on the face of the deep. And the Spirit of God was hovering over the face of the waters. Then God said, "Let there be light"; and there was light. And God saw the light, that it was good; and God divided the light from the darkness. God called the light Day, and the darkness He called Night. So the evening and the morning were the first day. (Genesis 1:2-5)

How would a simple, straightforward reading take this text? For one, the writer describes an ordinary day. The text identifies a light/dark cycle as a day, plus it specifies that the day spanned a morning and an evening. A child would know exactly what that means. I would have known what it meant, but by the time I first read it, I was already too educated against the possibility of miracles to allow them to explain beginnings. So, like countless people before me, I used the very cleverness that God miraculously created and placed into my mind to deny creation miracles, including the creation of my cleverness. Even my Bible teachers in children's church helped me justify any meaning for "day" in this text other than it being an ordinary day.

Long after that, when I got serious about finding answers, I began to see that arguments insisting that this text does not convey an ordinary day boil down to saying that the text does not mean what it says on its surface. And that opens the door to the bigger, more important question that this book attempts to answer: Does Genesis 1 or the Bible's own comments on Genesis 1 give enough reason to depart from "day" actually meaning an ordinary day? We will return to this later in more detail. I just want to set forth that without a bias against miracles, the text seems to clearly say that God created the earth in an ordinary day—and even a little kid gets that.

Another observation from this text that even the child will notice—whether consciously or not—is its tight sequence of events. One thing happens right after the other, almost like a giant run-on sentence. The earth was dark and the Spirit was hovering, then God spoke and immediately what He said happened, and God approved of that and divided light from dark. Whew. Without waiting beyond that first morning/evening cycle, God named the light Day and the dark Night.

What does Genesis say in its first chapter? At the start God made the light and dark cycle that today we still call "day." On the second plain old day, He placed some space between upper and lower waters. On the third, now down to Earth, He separated land from water, plus He made plants. The creation and placement of the sun, moon, and stars happened on Day 4. Now the new home for man and animals was almost ready. Days 5 and 6 return to Earth with air, sea, and land animals, and finally mankind. The whole first chapter portrays each action happening directly after the prior one. The Creator didn't rest until it was all done and very good.

In this first chapter, God commands into existence all the main features of the cosmos. "Beginning" initiates time itself. God made time. The newly formed "heavens" made space for Earth and all things physical to exist as material bodies with at least the three spatial dimensions of length, width, and height. Working backwards, the chapter described how God spoke into existence mankind, animals, lights for day and night, plants, land for them to grow on, water for plants plus water in a sea, the space for birds to fly and stars to shine, and the good old dependable cycle of day and night. It's all there in black and white, and it reads like a string of facts, as though someone actually witnessed the most stunning series of miraculous events in all history.

Who God Is

By this point, Genesis appears to have revealed perhaps the most widely known and basic definition of God. He is the One who made it all. He is the Creator. The Creator is God. More than that, the first verse in the Bible suggests key qualities of this Creator. For example, causes are greater than their effects. I played a role in causing my son to exist, but he will never be more human than me. Why not apply this principle to basic features of reality to show something of the nature of God?

Eternal

Consider the Bible's opening phrase, "In the beginning." The Creator of time must be greater than time. This implies that the God who created time must exist outside of time. We find God's existence—the essence of His being—outside of time itself. Similarly, the Creator of matter must be greater than matter. In other words, if God's very essence were physical, then He would have no more power to create physical realities out of nothing than mankind can invent new colors. Any God who can create space itself must exist outside of that which has length, width, and height. The God of Genesis 1:1 must be greater than that which He called into existence. He lives (for He created life) outside time, matter, and space. No wonder Scripture calls Him holy—a word that means set apart and like no other!

Miracles

Genesis sets forth a bold and rare assertion: a God who must have existed prior to this universe called the universe into existence. Now that's power. Psalm 33:9 says, "For He spoke, and it was done; He commanded, and it stood fast." He didn't have to consult a committee. He didn't need to spend hours in design meetings or eons refining assembly lines for the manufacturing of plants and animals like some cosmic North Pole elf. He just commanded it, and each piece, whether a material leg or the programmed coordination to use it, instantly became perfected reality.

The Bible opens with a jolt. It says God just made it. Is this an isolated assertion? Can we reasonably say that some ancient author must have been mistaken when he wrote that creation happened miraculously, or does the rest of the Bible agree with this sudden series of miracles in creation?

Digging deeper in the Bible reveals an answer. For one, Psalm 33 exactly matches this sudden miracle concept of creation. How did God, through the prophets and apostles, introduce Himself to those who did not know Him? As the Creator, of course. For example, God through the prophet Isaiah foretells a coming Servant with whom He will be pleased and upon whom He will put His Spirit. Then He says:

> Thus says God the LORD,
> Who created the heavens and stretched them out,
> Who spread forth the earth and that which comes from it,
> Who gives breath to the people on it,
> And spirit to those who walk on it:
> "I, the LORD, have called You in righteousness,

And will hold Your hand;
I will keep You and give You as a covenant to the people,
As a light to the Gentiles,
To open blind eyes,
To bring out prisoners from the prison,
Those who sit in darkness from the prison house.
I am the LORD, that is My name;
And My glory I will not give to another,
Nor My praise to carved images.
Behold, the former things have come to pass,
And new things I declare;
Before they spring forth I tell you of them." (Isaiah 42:5-9)

Who else is God but He who made the heavens, the earth, and people as bodies and souls? Many who would never admit to knowing Him as a personal friend at least know about Him. This passage says that the same God who made everything promises to give His Servant "as a light to the Gentiles." He even identifies Himself as the One who announces events before they occur—events like His Servant coming to Earth as a light of truth to the Gentiles (non-Jews).

If the ancient Jews concocted Genesis to establish that their God was better than the Gentiles' gods, then why does this creation assertion from Jewish literature make it sound like God is really interested in non-Jews? The One who wants to know me is the One who made me. If this same One was able to craft a garden as a place for Him and the first human couple to meet in the past, then He proved He will be able to craft a new and appropriately glorious place for Him to fellowship with us in the future.

Know-It-All

Also, a God with enough power and wisdom to create everything and who has an existence independent of the time and space of His creation must know everything, including all future events. So, we at least find logical consistency with a God who can create anything *and* tell the future. Plus, the very Servant that Isaiah 42:5-9 (quoted above) foretells matches the New Testament descriptions of Jesus Christ. Matthew 11:5 says about Jesus' earthly ministry, "The blind receive sight," just as Isaiah 42:7 foretold. Here we see that even the Old Testament identifies Jesus as the Creator.

The apostle Paul provides another example of introducing this Savior as Creator of those who have never heard about Him. They probably never read Genesis. Speaking to a crowd of Greeks, he cleverly identified with one of their pagan altars ascribed to "the unknown god." Paul said to those ancient Athenians:

> Therefore, the One whom you worship without knowing, Him I proclaim to you: God, who made the world and everything in it, since He is Lord of heaven and earth, does not dwell in temples made with hands. Nor is He worshiped with men's hands, as though He needed anything, since He gives to all life, breath, and all things. And He has made from one blood every nation of men to dwell on all the face of the earth, and has determined their preappointed times and the boundaries of their dwellings. (Acts 17:23-26)

Paul must be describing the same God as the one working in Genesis 1. Paul even identifies this Creator as having all knowledge of the future, just like Isaiah said. Again, God is getting introduced to folks who don't know Him as the One who made it all and knows it all.

Finally, a collection of 24 elders in heaven sing this specific song to God in His very presence according to the Revelation given to John:

> You are worthy, O Lord,
> To receive glory and honor and power;
> For You created all things,
> And by Your will they exist and were created.
> (Revelation 4:11)

Who is God? He has revealed Himself in many ways to many people, but Genesis says and the rest of the Bible agrees that He alone is the Creator of the universe.

Genesis, God, and Me

Compare this with what I first believed about God. I would call God my Savior, and for good reason—He saved me from the just penalty that my sins had earned! Who can forgive sins but the one against whom those sins were perpetrated? Thus, only God could forgive my guilt before God. I would and still agree that Jesus is the way and the truth and the life, and Jesus is God. But my God was all about forgiveness and salvation and had nothing to do with the origin of the universe, Earth, or mankind. After all, God as my Creator would confront evolution and a whole world full of scientists who assured me that science has proved evolution is a fact. They all sound so confident when they describe fossils or stars as ancient and evolved. They speak as though they witnessed the formation of that rock with its fossils millions of years ago or that galaxy and its stars billions of years ago.

I soon encountered a question. If I as a Christian could not stand on Christ as Creator, then was I really talking about the same Jesus as the One the Bible describes? I began to realize that my concept of God as Savior was good, but it was only a good beginning. God is more than a Savior.

As this concept dawned on me, I began to think through its consequences. With no God and no Creator, then what does sin mean? Sin becomes a farce. No wonder we have schools of psychological thought teaching that our once-disagreeable tendencies are just part of what evolution made into us. To them, sin does not exist. To them, as to the younger version of me, Jeremiah would have rightly said:

> "Were they ashamed when they had committed abomination?
> No! They were not at all ashamed;
> Nor did they know how to blush.
> Therefore they shall fall among those who fall;
> At the time I punish them,
> They shall be cast down," says the LORD. (Jeremiah 6:15)

I began to face these tough questions. If God made me, then what does He expect of me? Did He make me for some purpose? If God is really my Creator, then He may have a plan for me, both in this world and the next. We then bear some obligation to find out about it. I soon realized I should go ahead and swallow some humble pie. I began to evaluate my life's direction in light of this Creator.

Let me illustrate what I found. I used an online tool to make what some call a wordle. I input the text of Genesis 1–11 and let the program run. It assigns the largest sizes to the words most often repeated. This Genesis 1–11 wordle clearly shows what the text is all about—the Lord! Remove this text and we remove the very foundation for knowing who God is. I had lived a life that treated pre-Abraham chapters in Genesis with contempt. No wonder I encountered so many hurdles while getting to know Him.

Does Genesis matter? Ultimate origins make a definite difference. Genesis describes the first and possibly greatest miracle in the Bible. It asserts that God spoke all this into existence. And since the rest of the Bible agrees with this, we can say that the whole Bible identifies God as the Creator. Therefore, those who doubt the Genesis account of creation open the door to a shriveled view of the God that the Bible describes and that history and the world verifies.

Wordle generated online shows the more-often-repeated words from Genesis 1-11 in larger font sizes.

My Struggles with Genesis

Does the Genesis assertion of one miracle after another during creation in six days conflict with modern origins ideas that invoke ages of natural process? Absolutely. I mean, can any honest reader objectively pinpoint in Genesis or anywhere else in Scripture even a mere hint of long-acting processes like galaxy formation by eons of gravitational attraction or land creatures very gradually morphing into flying creatures? None at all.

When I first read Genesis, tension mounted between what it seemed to say and what I thought I knew to be true or scientific…or at the very least what scientists and culture at large believed to be true. At first, I wanted nothing to do with seeking a solution. It would take too much effort. Besides, it wouldn't be worth the labor anyway since science had settled the matter long ago. I also thought academic wrangles about origins had little impact on what happened in my day-to-day life. But like an oil tanker changing its heading, my mind slowly changed as I began to see the shallowness of these thoughts.

Tension between Genesis and scientific origins would not go away. My curiosity mounted until I just had to find some kind of answer, even if it meant we just can't know the truth about origins—that mankind is forever doomed to ignorance and must simply plow forward with whatever scraps we can stir into our origins stew. How unsatisfying, but so be it if necessary, I thought.

So, I began to probe. Does Genesis really mean what it seems to say on its surface? Do scientists really know that it didn't happen the way Genesis says? If so, then how did they find out? Which scientists did the work, and did those scientists bring any biases to their science? What experiments persuaded them to doubt the ancient text? Did they all behave like Spock on the original *Star Trek* TV shows—with that cold, calculating, and irrefutable logic?

I wanted to know the truth. So, I set my jaw in preparation for reading more about science, scientists, the Bible, and Genesis. Soon I came to Genesis 3.

Sin and Death

Genesis 3 revealed that the first, pristine creation soured. Adam and Eve disobeyed the only command from their Creator—to not eat from a particular tree. They had access to a cornucopia of delightful fruits at every turn. They had access to God Himself, who the text says personally, physically walked with them daily! And yet they sinned by choosing not to believe what He told them, by breaking His law. They instantly, and for the first time, felt shame. But God did not leave them in a lurch. He sought them out. When He found them, He cursed the devil, here called the "serpent" (Revelation 20:2), cursed Eve with pain in childbirth, cursed the very ground they stood upon, and then cursed Adam with perpetual toil. Not the best day.

God promised that if they ate the forbidden fruit, they would die.

Well, they did. Eventually. But, boy, did it take a long time. Apparently, Adam lived for 930 years. That does not happen today. When I first read this, it sounded like a bold affront to science if not to common sense. I have since changed my mind after considering the effect of genetic bottlenecking at the Flood on once perfect genes, as I explain in chapter 3. If those ancient authors faked their numbers, then how did they get them into a decay curve when plotted from Genesis 11 life span data that looks familiar to biologists?

Meanwhile, we must admit that God did promise death, and death came sure enough. It still meets all of us in the end. In other words, this process of getting the gist of Genesis kept leading me to admit that it seems to explain basic features of the real world, including the inevitability of death and even our universal sense of justice, as revealed in our reactions to those who injure us.

Judgment and Grace

Watery Justice

Fewer than two of Adam's lifetimes later, Noah and his sons built an ark. The Ark that God specified to Noah would have looked like a huge, elongated chest with three decks inside it. The Flood account in Genesis 6–9 leaves no doubt that its author specified certain details—especially those that we read two or even three times. For example, notice the death of "all" animals outside the Ark in each of three consecutive verses:

> And all flesh died that moved on the earth: birds and cattle and beasts and every creeping thing that creeps on the earth, and every man. All in whose nostrils was the breath of the spirit of life, all that was on the dry land, died. So He destroyed all living things which were on the face of the ground: both man and cattle, creeping thing and bird of the air. (Genesis 7:21-23)

In these verses, as in earlier Genesis chapters, the word "cattle" translates an original word that indicates domesticated animals. "Creeping things" are generally not domesticated but live in the wild. "Bird" refers to a broader "flying creature" and would include any flying animal, whether mammal, reptile, or bird. But here Genesis hammers the point that if they were not on the Ark, they died.

How did all those animals and men die? The text insists in triplicate (Genesis 7:17-20) that water covered the whole world. Why did man, animals, and indeed the entire surface of the earth suffer such a horrifying calamity? The Bible does not mince words. Man had reached such a low state of depravity "that every intent of the thoughts of his heart was only evil continually" (Genesis 6:5). Worldwide judgment became necessary to preserve the very last family who feared God amidst an entire world of people who reviled Him. Worldwide judgment became necessary to preserve the future, including the writer and reader of this very book. This path would reestablish hope for a future without the same degree of abject and universal wickedness. Remnants of God's originally good creation would survive long enough to rekindle hope that someday God's promise of a woman's Seed who would damage the head of the serpent would find fulfilment. Judgment became real.

I once asked a dear, unbelieving family member what she thought about judgment. In no uncertain terms, she called it a "crock." She totally rejects, even to this day as far as I know, the very concept of God judging men like a just judge would and should do. Now, it's one thing to disagree academically over the concept of judgment, but she went beyond that. She was emotional in her disdain toward any kind of judgment. Why? To admit the possibility of judgment is to admit that there is sin that needs to be judged. To admit sin is to admit guilt. And that requires humility. So, a proud person wills herself to pretend that sin is "religious" or even "superstitious" instead of real. Proud people refuse to admit sin, guilt, or judgment. How can this bias fail to affect the way one views Genesis? Scientists are people, too. They have biases. Even a scientist's bias against sin and judgment can affect her view of Genesis.

Did God really judge the whole world? Peter believed it. He urged his readers to cling so tightly to the Flood narrative described in "the words which were spoken before by the holy prophets" (2 Peter 3:2) that it would inspire us with confidence "that the heavens and the earth which are now preserved by the same word, are reserved for fire until the day of judgment and perdition of ungodly men" (2 Peter 3:7).

Therefore we ought to conduct ourselves with holiness and godliness.

Genesis and Salvation

Does Genesis matter? Well, if God really did judge the world with water according to the Genesis record, and if this same God promises to judge the world again by fire, then the one who rejects the record of the first judgment will probably doubt the promise of the second. After all, what we learn about both judgments comes from the same Bible. Genesis 3 clarifies God's judgment by God's justice in sentencing the serpent, Eve, earth, and Adam. Genesis reinforces the historical reality of universal judgment when God cleansed the globe with water during Noah's life.

If we are to take these Genesis judgments as some kind of fable, metaphor, or poem rather than history, then why not take the promises of future judgment that way too? And if God's judgment of mankind is a "crock," then so must be God's placement of our judgment upon His Son, Jesus, "being found in appearance as a man, He humbled Himself and became obedient to the point of death, even the death of the cross" (Philippians 2:8). No wonder modern conventional thinkers, including a former version of myself plus some of my family and friends, have no idea why they need salvation. Those who reject the very possibility of divine justice would reject judgment accounts like those in Genesis.

So far, I've divulged what I think is a plain reading of Genesis' depiction of who God is (He is the Creator), what sin is (breaking God's revealed law), why death exists (as the just penalty for sin), and the significance of judgments (God takes disobedience very seriously). But Genesis presents another first: grace!

"But Noah found grace in the eyes of the LORD" (Genesis 6:8). What is grace? Typical Bible teaching says it could either mean "undeserved favor" or "divine enablement." The first definition definitely applies to Noah. God favored Noah. Why would He favor Noah instead of the others on Earth at that time? Probably because the others wanted "only evil continually" whereas Noah wanted God more than he wanted wickedness. "Noah walked with God" (Genesis 6:9). Can anyone walk in harmony alongside someone they despise? Noah feared, honored, and trusted God. "By faith Noah, being divinely warned of things not yet seen, moved with godly fear, prepared an ark for the saving of his household" (Hebrews 11:7). He set an example of faith and godly fear that many men and women who walked with God followed throughout biblical history up until today.

In this chapter I tried to take a fresh look at what Genesis says. That way we have a clearer basis for what's at stake for those who reject it versus what each may gain from taking Genesis at face value. Along the way, I tried to tell the beginnings of my story. I went from holding Genesis at arm's length to embracing the book. One of the first steps in my transition involved looking more closely at what Genesis actually says. I did not find in Genesis some dusty historical record with stale or tedious accounts. Nor did I find confused ramblings of ignorant shepherds. Genesis talks of a real Adam from whom all men descend, a real garden where Adam walked with God, and a real problem of rebellion, not fanciful tales about various gods' whimsies like we read in the ancient Near Eastern creation myths. The pages came alive with plain statements about who God is and who I am. I saw suggestions about what I was made for and how I might move toward real purpose in life.

But a huge hurdle stood in the way of me believing the beginning statements of Scripture: evolution. Evolution from hydrogens to humans over billions of years is established fact, right? The rendition of history that Genesis reveals frankly flies in the face of the evolutionary history I learned and even taught. I liked them both! I liked the clarity and confidence in what Genesis said, but I liked the unity among what the world's scientists said. It didn't matter what I liked. I would believe whatever was true.

So, which one is true? Or could they both be partly true? The next chapter unfolds what I found about that.

2
GENESIS AND EVOLUTION

"Now that we know that Adam and Eve never were real people the central myth of Christianity is destroyed. If there never was an Adam and Eve there never was original sin. If there never was an original sin there's no need of salvation...I think evolution is absolutely the death knell of Christianity."[1]
— *Frank Zindler, 1996*

To Contradict or Not to Contradict

I'm going to start this chapter with a story that brings up a principle. Then I'll try to apply that principle to the way we handle Genesis.

Here's the scene. About a dozen young adults sat in a classroom discussing Scripture. Math posters hung on the walls for use in our Christian school Monday through Friday, but this older group squeezed into high school desks, ignoring room décor during Sunday school. We read a Bible verse together, then the teacher opened the floor for comments. One young mother shared her thoughts and included the phrase "Perception is reality." That sent a red flag up my mental pole.

How can we *know* the objective reality that "perception is reality" if something as subjective as perception equals reality?

These are the things I think about. I know it's weird, but I came out of the oven this way. Sometimes I wish I could sling phrases around without thinking much about them like this young mother did. I think I know what she meant. Something can be real or true without us knowing it. Then it becomes real *to us* when we become aware of it. But that's not what the expression says. Its grammar suggests that reality becomes *universal truth* only when we become aware of it.

I couldn't concentrate on the rest of her response for want of resolution to this rankling quirk. I deliberated, slow thinker that I am, for hours. Perception is reality. Really? If so, is that statement itself reality or just perception? I called a friend who had training in philosophy. Thankfully, he helped me sort it out. He said it boiled down to a contradiction.

On the one hand, strictly speaking, the expression makes a truth claim. It claims that reality depends on perception. But on the other hand, the expression assumes that reality does not depend on perception. It assumes a universal, objective truth that reality is subjective perception. It says one outward "truth" while assuming an opposite truth. No wonder that red flag blinded my mind.

Eventually, I felt comforted that the red flag wafted across my cortex for a real reason. But that got me thinking about contradictions in general. A square circle. A black white. A false truth. 2 + 2 = 21. Such things do not exist. Reality does exist, for denying it assumes it. Could I deny Genesis so easily, or would that involve similarly schizophrenic contradictory concepts? Does evolution contradict Genesis, and do either of those contradict truth?

Side by Side

Some say maybe God used evolution to create the world. I have heard, and at certain times believed, an array of attempts to merge evolution with Genesis. I eventually found them all poorly thought out. Careful comparison between the doctrines of evolution and the teachings of Genesis led me to conclude that the two views not only mix like oil and water, but they also make exactly opposite claims.

For example, consider what Genesis 1:24 says about the creation of Day 6 animals. "Then God said, 'Let the earth bring forth the living creature according to its kind: cattle and creeping thing and beast of the earth, each according to its kind'; and it was so." It begins with "then God said," emphasizing the direct action of God's miraculous power. Evolution teaches that natural—not supernatural—processes brought forth the animals.

Next, the verse says that God used the stuff of earth to bring forth the living creatures. Evolution teaches that the first animals evolved not from earth but in the sea, and land animals emerged from sea life. Genesis 1 reiterates 10 times that God made creatures to replicate themselves according to each creature's kind (basic and reproducible body plan). Evolution teaches exactly the opposite. It says that all of today's creatures emerged from totally different creature kinds. Genesis says that God made either fish or cat, ape or man right from the start. Evolution teaches that fish became cats and apes…and men.

Don't believe me? Then just allow me to quote a college biology textbook. "We belong to a highly specialized group of bony fishes…we are fishes whether we like it or not."[2] The end of Genesis 1:24 says, "And it was so," meaning that the very moment God spoke the command, it was done—right then on Day 6. Evolution teaches that animals morphed from one to the next over untold eons.

If you're keeping count, that totals five opposite teachings in just one verse. It would appear that Genesis and evolution fit together about as well as dark and light.

The order of events given in Genesis also clashes with what I increasingly regard as the man-made tale of evolutionary origins. According to evolution, stars formed first, then the sun, then a molten earth, then water somehow reached the earth after it cooled. Plants and animals followed, with birds arriving after land animals. I wonder if God created the features and creatures in His cosmos in the order described in Genesis to thwart in advance these man-made naturalistic origins stories. Scripture says watery Earth came first, then the sun on Day 4. It asserts that birds flew before land animals ran, whereas evolutionary museums teach that land animals (theropod dinosaurs) evolved into birds over millions of years. Never mind about fossils of the bird *Protoavis* found far below dinosaur rock layers among the deepest-buried land animals in all the fossil layers. Plenty of similar details fall in the wrong order. Again, this suggests one ought to take either the words of naturalists—who by definition will not take God's Word for what happened—or the words of God as true.

We should each ask ourselves which history seems most reasonable.

Or if we are unable to study the contrasting historical scenarios of evolution and creation, then we should ask ourselves on whose authority do we accept one versus the other and why. This exercise might expose, for example, that we have chosen to trust what most people around us believe (i.e., evolution) simply because we are too cowardly to bear the ridicule that we fear they may heap upon us or simply because we would rather choose a worldview that makes our sin feel more comfortable. Neither rationale is scientific or helpful. But talking it through and acknowledging our motivation surely is a step in the right direction. For me, the choice to believe the truth wherever it is found became both more important than succumbing to peer pressure and an easier choice given the science that supports Scripture.

Cut and Paste

It's kinda fun to cut and paste. It began in kindergarten. We would cut out shapes of paper with our mini scissors and paste our art onto a crude canvas. We made one-of-a-kind masterpieces. Creativity blossomed. When I use my computer to cut text from one place and paste it in another, I can feel like I'm in control. I get to choose what goes where. That's all fine and dandy as long as I submit those cut and paste decisions to the Lord—dedicating creative decisions to His care and honor. But do any of us have the right to cut and paste God's words?

If I add my words to God's, then I put myself in God's place. If I subtract from them because I doubt them, then do I not in some way also doubt God, who claims to inspire Scripture's every word? Maybe we question His judgment, wisdom, or intention. Maybe we just want to be in charge. Whichever flavor of tension we experience, the words of Genesis, and thus the words of all the Bible that leans on that seminal document of origins, can force us to investigate what may be happening in our hearts between us and God.

This was the origin of the first sin and remains the basis of sin to this day. The serpent, who was the devil, asked Eve, "Has God indeed said, 'You shall not eat of every tree of the garden?'" (Genesis 3:1). Truly, today conventional influences like movies and college classes inject similar doubts that echo the serpent's "has God indeed said" question.

Has God indeed said six days? Maybe that's not what He meant.

Has God indeed said creation and not evolution? Maybe He meant evolution. Has God indeed said that everything was "very good" in the beginning? Maybe He meant everything except all the animals (that died for billions of supposed years and left their fossil remains before sin began). Has God indeed said that He raised Jesus from the dead? Maybe the apostles all imagined the resurrection or concocted the story in collusion. And so on.

I quickly came to realize that I needed to know what I believed and why I believed it. My studies led me back to Genesis. Here's the rough progression of my thoughts as I worked through them over several years. I tried over and again to cut and paste this or that concept into and out of the Genesis narrative. My pasted platitudes, when interspersed with those old words, held all the power of paper bullets. The dissonant merging of contradictory concepts like eons for days and natural with supernatural processes left me perpetually dissatisfied. Beside that, the whole approach smelled funny. Who was I to presume I could paste ideas into the Bible? I had to find out if I really knew enough or if I or scientists really had the right to adjust Genesis.

Day-Age Theory

I remember it like it was last month, but the conversation happened 30 or so years ago now. I was riding shotgun, heading west toward my East Texas college campus in my friend's hand-me-down pickup truck. We began discussing origins. He seemed a confident fellow. With his left hand on the wheel, he raised his right index finger to make a sure point. He said that a verse in the Bible says a day with the Lord is as 1,000 years. Further, this definition of day applied to the creation days in Genesis. He soon slid in the idea that *day* could mean not just 1,000 years but any great length of time. So, he said, who can know how many ages elapsed during that creation "week"?

I rubbed my chin and nodded assent. The proposal had a tempting appeal. By treating the words of Genesis 1 in this unique way—and after all, it does describe a unique time in history—I could squeeze at least 6,000 years if not millions into the six creation days. That way I could avoid apparent conflict between the "science" of eons and the "history" in Genesis. Tempting as all that felt, the conviction soon slacked.

I later reread the first chapter of Genesis. I had a hard time circumnavigating its clarity. "Morning and evening, the second day," it said. What kind of morning could dawn on a 1,000-or-more-year eon? None that I could imagine. Then I found the verse about a day being as 1,000 years in 2 Peter 3. I discovered that my pal neglected to quote the second part of that same verse. It says, "And a thousand years are as a day." So…what? Am I now supposed to treat every 1,000-year span as a mere day? Clearly the verse was not defining Genesis 1 days or any days. The misapplied verse in 2 Peter simply says that God lives outside of time. Of course He does—He is the eternal Being, not time-bound temporal beings like us. Meanwhile, Genesis defines its own days with words like "morning" and "evening."

Soon I learned that day-age theory is the name of my friend's proposal. At first blush, day-age had its appeal. But it ended up clashing with the text. Unless another interpretive option emerged, I would still be forced to choose to either doubt God's Word about the timing of creation in favor of scientific consensus or doubt what we know from science in favor of this ancient book. Another interpretive possibility soon slid into the gap in my confidence.

Gap Theory

"In the beginning, God created the heavens and the earth." (Now wait billions of years.) "Now the earth was formless and void." That's the gap theory.

And like day-age, I liked it at first. It offered an easy way to reconcile a version of creation with my assumption that the universe had existed billions of years. Just jam those eons in between the first two verses. Although I did notice that those supposed eons seemed to come from outside the Bible instead of from within the text, I figured that time was allowable.

I read about big flaws with the gap theory not long after that. For example, the whole reason for squeezing ages into the text is to accommodate millions of years of fossil deposition. The logic looked like this: Thousands of feet of rock layers with fossils take millions of years to form. Millions of years (i.e., deep time) must therefore be made to fit into the history of the world. The position within Genesis 1 that can absorb them with the fewest collateral damages to the text occurs right before creation week Day 1. Therefore, the millions of years happened before God re-created the heavens and Earth's surface. The more I thought and read about it, the more collateral damage I began to see. For example, placing fossils before the creation week puts death before sin, and that raises questions about the whole reason for Christ's death. Why would He need to die to pay for our sins if death is not the penalty for sin after all? Out with the gap. In with the same old-time dilemma.

Progressive Creation

Progressive creation teaches that God created one thing, waited eons before creating the next thing, and so on—adding each creature or feature to His cosmos in basically the order that atheistic evolutionists teach. Accordingly, God eventually took two of his ape-like humanoids and made them into Adam and Eve by placing souls inside them.

By the time I learned about this, I was already leery of attempts to accommodate evolutionary time into the Genesis record. It didn't take me long to decide against progressive creation since it suffers the same problems as the day-age and gap explanations. Like day-age, it forces the reader to insist that the text means something other than what the words say. For example, progressive creation has the sun formed first and then Earth. It also insists that "day" cannot possibly mean a day in the normal sense. Also, like gap it introduces death before sin.

After running through day-age, gap, and progressive creation, I started to see a trend. Each of these proposed solutions modifies the Bible while leaving the scientific consensus of billions of years unchallenged. I couldn't seem to find a crack in the armor of God's Word, so I began to think more in terms of seeing how infallible the scientists' notions were. We will breeze over some of the fruits of that emphasis in a later chapter, but for now I will admit that I ended up finding the consensus position looking like Hans Christian Andersen's emperor with no clothes. The more I looked, the fewer faults I found in Genesis and the more faults I found in scientific dating methods. The trend reached a point where my dial turned completely around until all the once-intimidating science lost all its fearsome teeth. Nowadays I refer to day-age, gap, and the like as compromise positions because all of them compromise God's Word, and none of them challenge man's word about our ultimate origins.

I have lived long enough now to also notice trends among Bible academics. Gap theory's popularity waxed in the mid-20th century but has since waned due to its lack of defensibility. But still feeling the hot breath of science-sounding insistence on billions of years, some thinkers have lately concocted a new compromise.

Exalted Prose

Some influential evangelical authors now teach that Genesis belongs to a unique literary genre of ancient Near Eastern (ANE) texts that some have termed *exalted prose* or *mytho-history*.[3] Categories like these come not from ANE scholars but from philosophers and theologians who have all admitted their belief that science has proven a history that the Bible does not convey. Placing Genesis 1 in imaginary categories like these would mean that it was never written to be read plainly. As a result, God did not create the universe in six plain days. Some extreme views purport that Genesis 1 was never meant to describe details of an actual creation week at all. What does this imply, and where did it come from?

These Bible college and seminary teachers assert that the original author of Genesis 1 patterned his words after those found in ANE texts like *Enuma Elish* from ancient Babylon. They assert that ANE literature is the key to a proper understanding of Genesis. As such, that ancient human author of Genesis 1 never meant to convey history. Further, those students who study ANE writings would surely learn that the human author of Genesis 1 did not intend to describe events but instead to exalt the Hebrew God above the surrounding pagan gods. These teachers refer to the Adam story, the Flood story, and the exodus story as mythicized.

How does a god exalt himself by inspiring a divine text that merely lays out a false history?

The doctrine of inspiration comes from the pages of Scripture itself. It teaches that a plain-thinking reader of just the Bible (for example, Jeremiah 7:1, Proverb 30:5, Luke 24:27, 2 Peter 1:19-21, Hebrews 1:1-2, and most of all 2 Timothy 3:16) and not the words of scientists, scholars, or ANE texts can clearly see that *God used fallible humans to write down His infallible words*.[4] This being so, why would a god who allows mix-ups to muddy his message earn added exaltation rather than exasperation? Why would a god who fails to offer a single clue to whole generations of readers as to why he would bury what he really wanted to say beneath words that describe historical events deserve exaltation rather than excoriation? One wonders why any evangelical would buy these ANE-first books.

New Zealand-based educator and author Michael Drake aptly critiqued this whole ANE push in his helpful book *The Misted Worlds of Genesis 1*.[5] In it, he points out that none of the ANE texts even *address* the very beginning! Those stylized fairy tales talk of grumpy gods who reorganize preexisting material. In contrast, Genesis stands alone in its bold and plain description of a transcendent (and quite content) God who calls into existence that which had no prior substance.[6] The rest of the Bible agrees with its own first book, of course. Consider Hebrews 11:3, "By faith we understand that the worlds were framed by the word of God, so that the things which are seen were not made of things which are visible." In the end, these ANE-first scholars describe no actual ANE texts that fit the fictional category of exalted prose. Much like the "scientific" category of exobiology (the search for life outside Earth), these researchers have no actual examples to examine.

There remains no such thing as exalted prose.

Where are the exalted prose texts? Only Genesis 1 ends up in this supposed bin. Though wrapped in sophisticated terms, in the end this idea reduces to another failure to give "the words their normal meaning in their normal context"[7] so as to make Genesis say something that sounds more palatable to pagan demands for evolutionary time. Imagine interpreting a medical doctor's prescription in this way so that it no longer means what it says. That could kill you. But using this approach with the Bible carries even more dire consequences, since it contains the prescription not just for this brief physical reality but for an everlasting afterlife. In the end, the only objective way to decide what the Bible's text or any other text means is to give its words their normal meaning in their normal context like we do in normal conversation every day.

Do Bible readers benefit from paying attention to genre? Of course! Drake wrote:

> Genre cannot be ignored: poetry is different from letters which are different from histories. But having said that, applying the intricacies and uncertainties of complex genre studies to Bible reading can distort that reading by conforming the text to non-biblical patterns of uninspired literature.[8]

By conforming Genesis 1 to not only non-biblical patterns but non-real patterns supposedly from ANE literature distorts that reading by erasing the miraculous power of God to speak time, space, and matter into being.

What should be said about those who admonish the Church to rely on their unique notion of pagan culture rather than on God for deciding what Genesis *indeed* says (Genesis 3:1)? Drake noted that ultimately they make "the reading of Scripture subject to the mysteries and vagaries of specialised scholarship; that effectively makes the Bible inaccessible to ordinary readers unless mediated by an elite magisterium of biblical scholars."[9] Heroes died during the Reformation to wrest the Bible from the priests' illegitimate grip. Today's elites now wreck the Reformers' gains by cloaking Genesis beneath tangled webs with fancy labels around Genesis that include calling it exalted prose. They might end up exalting only themselves.

What does the ANE stance have in common with gap, progressive creation, or other compromise positions? They all take for granted that science has proven billions of years. I needed to hear more about the problems with old-age dating techniques—problems that I never really investigated before. This quest led me to discover that deep time did not have its origins in science but in the self-named Enlightenment thinkers of a few centuries ago.

Free Science from Moses

The comparison between Genesis and evolution in this chapter would benefit from a peek into the origins of two key evolutionary concepts: deep time and infinite creature plasticity. Did data from observation or experiments demand that scientists adopt deep-time dogma and fish-to-philosopher philosophies? Can any creature morph into any other creature by natural processes alone? Could someone have crafted these concepts in such a way that they merely pose as science?

Starting in France

Straw man arguments began the break from medieval respect for the biblical Flood. In this all-too-common break with logic, opponents of a view attack an imagined (typically extreme) version of that view instead of the real view. Combatants claim victory over their opponents when all they really did was beat up a flimsy opponent of their own making.

In 1720, René Réaumur studied sedimentary layers near Tours, France. He found fossil leaves mixed with fossil shell fragments. He said the leaves were "laid too neatly to be attributable to such a violent event as the Noahic cataclysm."[10] Réaumur portrayed a Noahic Flood that during its year-long upheavals could not involve mud flow rates gentle enough to lay down leaves without obliterating them. But the real Flood, involving pulses and surges of various magnitudes and directions from tsunamis to slack waters here and there, month after month, may well have done it.

Similarly, from 1749 to 1789, Comte de Buffon wrote his 36-volume work *Histoire Naturelle*. It featured imaginary vast ages for Earth long before scientists invented radioisotope dating. The very phrase *natural history* seems to supplant eyewitness history with modern interpretations of clues in rock layers. Buffon mentioned Noah's Flood, but his straw man version was too gentle to disturb trees and plants, let alone Earth's rocks and crust.

Since today's local floods can rip up rock, a worldwide flood would obliterate Earth's original crust. Since the Flood lasted over a year, it is reasonable to expect water and mud speeds to wax and wane at different places and times. Buffon wanted a flood too placid to leave a trace. Réaumur wanted one too violent to deposit leaves. Either way, their warped, straw man versions of Noah's Flood became much easier to ignore in favor of their preferred eons.

This long-age view still strangles minds today. It keeps many people from trusting the Bible. In some cases, that lack of trust extends far beyond questions about creation. It causes some to even question the

Bible's accuracy about Jesus. Did He really rise from the dead? Are miracles possible? Is there a God?

Key early naturalists leveraged their professional platforms to emphasize cherry-picked talking points. Even before Buffon used his high position this way, Bernard Le Bouyer de Fontenelle did the same. He became secretary of the Academy of Sciences in Paris in 1697. For over 40 years Fontenelle's responsibilities included summarizing the most noteworthy scientific research for the academy publications. He got to choose what was noteworthy. What did he choose? Whatever agreed with his low view of Genesis—for example, an old earth. British meteorologist and creation researcher Andrew Sibley summarized Fontenelle's endeavors in the *Journal of Creation*. He noted that Fontenelle "used the occasion to try and persuade French academic society that an ancient history of the world could be arrived at through studies of nature with very little regard for belief in Noah's Flood."[11] Anti-Flood sentiment then migrated across the English Channel.

Subtle Side Attacks in England

Famed philosopher David Hume borrowed Eastern religious beliefs like Hinduism to argue against deism—belief in a god who started it all but has stayed aloof since then. Deism denies miracles. Hindus believe in vast time cycles instead of the Bible's record of a recent creation. Hume expressed his deep-time leanings non-confrontationally through fictional characters in dialogue. In one of his books, Hume's character named Philo says, "The world, say I, resembles an animal; therefore, it is an animal, therefore it arose from generation."[12]

Hume knew Erasmus Darwin, Charles Darwin's grandfather. Erasmus wrote in *Zoonomia*:

> Hume…places the powers of generation much above those of our boasted reason; and…he concludes the world itself might…have been gradually produced from very small beginnings, increasing by the activity of its inherent principles, rather than by a sudden evolution of the whole by the Almighty fiat.[13]

But on what scientific basis did grandpa Erasmus and Hume base this particular story? Charles would cooperate in an even more subtle strategy to solidify that story into a new history.

Scottish lawyer Charles Lyell wrote a personal letter in which he suggested ways that even religionists could "join us in despising both the ancient and modern physico-theologians."[14] Like today's creation-believing scientists, the physico-theologians believed that Genesis events like the Flood left physical signs on Earth such as vast rock layers studded with fossils.

How would Lyell and his cohorts achieve their goal of garnering Flood despisers? In short, charm. Lyell wrote in a letter to George Scrope, "I conceived the idea five or six years ago, that if ever the Mosaic geology could be set down without giving offence, it would be in an historic sketch."[15] His plan came not from a desire to do good science but to "despise" and "set down" Genesis history. Creation researcher Michael Oard recently wrote about Lyell, "The strategy worked; the church and the culture were blindsided."[16] Charles Darwin advanced Lyell's goal with his wildly popular book *On the Origin of Species*. It spun an anti-creation "historic sketch" that masqueraded as science. These men's stories assumed deep time but never defended it.

In a letter to his son George, Darwin divulged his role in the plot to overthrow the Genesis Flood. He wrote:

> Lyell is most firmly convinced that he has shaken the faith in the Deluge & etc. more efficiently by never having said a word against the Bible, than if he had acted otherwise…P.S., I have lately read Morley's *Life of Voltaire* & he insists strongly that direct attacks on Christianity…produce little permanent effect: real good seems only to follow from slow and silent side attacks.[17]

What kind of world did these conspirators craft? A world in which the very people God created deny their own creation; one in which people daily drive over, walk on, and sometimes dig through Flood-deposited sediments all the while denying that the Flood even happened. Men like Lyell and his collaborators used straw man caricatures of the Flood, cherry-picked talking points that ignore the Bible, leveraged positions of influence, and planned stories written with charming subtlety that all gained the name of "science." Meanwhile, creation best explains why we have a world fit for life, and the Flood accounts for the billions of fossils buried inside countless tons of water-deposited rocks.

Genesis Without Evolution

I discovered that if I held on to evolution, I had to keep holding Genesis at a distance. Genesis and evolution make opposite truth claims. It's one or the other. After some time of asking questions and digging for answers, I finally began to lean toward the idea that Genesis deserved much more respect than I had originally thought. I just needed to know what to do with those millions of years that felt unassailable.

Looking back on my creation conversion, it seems that Noah's Flood made the biggest difference. It explained what I saw in the real ground better than any of the evolutionary ideas I had been told. The Flood explains why Earth's rock layers continue flat for mile after mile. Major rock units spread across hundreds or thousands of square miles, and they do that across every continent. Since each one comprises the same rock type with similar fossils, each such layer represents a single deposit. None of today's processes come remotely close to moving that much mud all at once. A worldwide Flood explains world-covering rock layers.[18]

The biblical Flood—as opposed to cultural flood legends—took 371 days to complete. It seemed to me that covering Earth with water for that long would provide the power to stack sediments like we see them. It seemed outlandish at first, mostly because of the ridicule that our culture pours onto Noah's Flood, calling it a fable or a mere child's story. But once I began to think in those terms—that a flood really could have covered the entire earth during the time of Noah—I saw that one year of depositing fossils left no time for evolution. The Flood started to erode my confidence in evolution, but I still had a long way to go. What about all that science that proves fish evolved into amphibians, apes became men, chemicals became the first cells, and that stars have cycled from gas to flame and back again? What about all the science that shows this and much more happened over untold eons? I had to discover—daunting as it seemed at the time—how defensible all this science really was.

3
SCIENCE THAT SUPPORTS GENESIS

> *"I conceived the idea five or six years ago, that if ever the Mosaic [Moses] geology could be set down without giving offence, it would be in an historic sketch."*[1]
> — *Charles Lyell, 1881*

The Power and Weakness of Science

In one of my favorite sermons, Pastor Voddie (pronounced Voe-dee) Baucham insightfully and entertainingly informs his audience of the limits of science. I know—he's not a scientist. But I am—or at least I have a Ph.D. in a science field, and I have conducted experiments for years and published peer-reviewed technical papers on them, for what that's worth—and I confirm wholeheartedly that Voddie has a better handle on the true purpose, power, and province of science than probably most scientists. We tend to elevate the impacts of our findings higher than they deserve. In his sermon titled "Why I Believe the Bible," viewable in several versions on YouTube, Voddie says that science cannot directly investigate the past. He's right.

And yet our ears get saturated with declarations about history based on "science," as though scientists have time machines in their laboratories. As one example among countless others, consider this line from a Cornell University news article first published in 2005 and stated oh, so many times before and since: "Chimpanzees and humans share a common ancestor, and even today 99 percent of the two species' DNA is identical."[2] It says that in the past, chimps and humans descended from a single ancestor. How do we know? Because of DNA similarity. Assuming we have accurate DNA sequences, reliable programs to compare them, and that evolution is true, we should expect to find similar sequences today.

But which scientist was there to witness any ancestor that chimps and humans may have shared long ago? Nobody. Which repeatable experiment showed that supposed ape-like ancestors diverged into two separate populations or that one of those populations stayed ape and the other became human as the story goes? None. Historical, past-tense statements like these do not come from observations. They therefore are not as scientific as their authors think.

And it turns out that the 99% identity line is false after all. Humans have three billion base pairs, whereas chimps have 3.2 billion. That's only a 94% identity right on the surface.

Plus, the technology to compare DNA base-for-base across different species didn't even exist in the 1970s when the 99% identity began its life as an evolutionary icon. The technology does exist today, and both conventional and Genesis-friendly geneticists independently arrived at the same conclusion of only an 84% identity.[3] Thus, the statement in the Cornell University news article suffers serious shortcomings. First, it repeats a fake fact, but more importantly, it represents countless other media and classroom statements that assume empirical science has the power to peer into the past. It can only analyze today's DNA. Every science-based statement about the past relies on inference. It could—and often does—get replaced by any number of other inferences that explain the same modern observations.

Voddie's insight about the limits of science means that when we read science articles that talk about the past, we can be certain that such "news" does not come from experiments. It comes from somewhere else. Too often we fallible humans with our strong biases against God and against His Word select "news" to fit our preconceptions—then invoke the authority of science to ratify our preference.

Empirical science involves systematic investigation of repeatable phenomena. It involves observing and recording. Its techniques answer present-tense questions like "How does this work?" or "What is this made of?" This kind of science provides the knowledge that inventors can use. Without empirical science, U.S. astronauts would not have landed on the moon (which they did six times), and we would not have cars or cell phones. Without empirical science, we would not have been able to observe bird flight. The Wright brothers famously used these observations when they modeled the first airplanes, which became the technological precursors of space flight. Without empirical science, Bible-believing founder of astronomy Johannes Kepler would have had a tiny view of God. He would never have reflected upon his discoveries as thinking God's thoughts after Him. When he measured motions in the heavens—empirically—he saw the Maker's genius, providence, and immensity. All this is great, but none of it examines the past.

Pastor Voddie described those who glibly say they don't believe in God or the Bible because science has not proven either one. He would ask, "Do you not realize that you don't use the scientific method to answer questions of history?" We *can* investigate the past, of course. But not with the scientific method only—or even mostly—and therefore not with empirical science. Historical science uses different tools in necessarily dim attempts to reconstruct possible pasts.

We see helpful illustrations of how historical science works when we attend law courts. There, judges and juries examine the past by evaluating reconstructed scenarios against the details that defense attorneys and prosecutors reveal. Judges and juries need to select the scenario that fits the most facts with the fewest gaps when lives are on the line. They sift through evidence. For example, they hear testimonies from reliable eyewitnesses. Attorneys can produce indirect evidence, too—also called circumstantial evidence.

Circumstantial evidence refers to facts or features that multiple scenarios can explain, albeit with more or less ease. For example, consider a shirt with the murder victim's DNA found at the house of the accused. Without more evidence such as video camera footage or direct testimony, honest juries are left to forever guess about how that shirt could have landed there. After all, someone other than the accused could have brought the shirt there. What does all this have to do with Genesis? Let's consider an example from archaeology.

Science and the Walls of Jericho

The archaeology of Jericho illustrates how shaky science becomes when researchers isolate their historical reconstructions to circumstantial evidence and ignore eyewitness accounts like those given in the Bible. The late, meticulous archaeologist Kathleen Kenyon's excavations of Jericho over several seasons included some radiocarbon dating results. Her team dated objects in Jericho's destruction layer. The resulting age estimates did not exactly match the biblical date for the destruction of Jericho as described in Joshua 6. Based on this and other reasoning, she concluded in the 1950s that Jericho was destroyed around 1560 BC, whereas the Bible's chronology pinpoints the event later—to the spring of 1406 BC.[4] She concluded that no city existed there at the time the Bible says. Those eager to see science dispute the Bible found her conclusion appealing. Who is right? Kenyon or the Bible?

Without a time machine in which researchers could travel to see Jericho's demise as often as they want, empirical science cannot directly answer this key question. But it can help. Later scholarship found abundant circumstantial evidence to correct Kenyon's age estimates. Before I describe that newer evidence,

Tel Jericho from cable car

however, why not review some of the limitations of dating techniques like radiocarbon dating? I'm about to describe some of the thinking that I wish I knew back when I was somehow led to believe that age-dating techniques were fully scientific and had no room for doubt. Once I started using them in my research, I found plenty of reasons to doubt them.

A radiocarbon "date" represents an age estimate. It's only as good as the assumptions that go into it. This means that it can give dates much less accurate to a calendar age than the impressive precision of its present-day measurements. For this technique, scientists measure carbon isotope ratios, not ages. To generate "ages," we plug those ratios into formulas with unknowable variables.

They're unknowable because we don't have time machines. Who knows the starting ratio of isotopes in the sample or in the atmosphere at the time of deposition? Many factors other than radioisotope decay over time can affect isotope ratios. Therefore, radiocarbon dating builds mere circumstantial evidence—details that more than one scenario can account for. We know carbon ages are circumstantial because new "dates" with new stories often sweep old "dates" and scenarios away, for example with Viking bones.[5]

What if the animal or human remains being tested ate a diet that was enriched in one isotope more than another, like the Vikings (see reference above)? What if a particular chemistry of rainwater dissolved and removed—or added—one isotope more than another during the millennia since the artifact first fell to the ground? What if solar radiation was different back then so that the upper atmosphere was not generating as much radiocarbon? What if the ratio of carbon isotopes had not yet reached a balance since God created the world or since the Flood rearranged the globe? Researchers must assume answers to all these and many other, unknowable factors.

Juglet from a Jericho tomb, Middle Bronze II (1800–1550 BC), displayed at the Michael C. Carlos Museum, Emory University

Additional circumstantial evidence would help clear up confusion. Fortunately for the archaeology of Jericho, pottery supplies age information for some archaeological settings. Unfortunately for Kenyon and those who cling to her conclusions today, the pottery tells a different story than her few carbon results.

In 1990, archaeologist Bryant Wood presented an array of circumstantial evidence that confirmed a roughly 1400 BC date for the destruction of Jericho.⁶ Bryant examined pottery in Jericho from the Late Bronze Age, which corresponds to the time of the exodus and conquest when Moses led his people out of Egypt and then Joshua followed God's leadership into the land that would become Israel. Why should anyone consider Byrant's circumstantial evidence superior to Kenyon's? Because he found much more than just a few broken pottery sherds. The evidence for a biblical date mounted to what a lawyer willing to look at all the facts instead of just those that support a particular bias would call "beyond reasonable doubt." Consider some of what his team found.

Bryant characterized telltale black and red-striped pottery that Kenyon and others dug up and handled as typical of Canaanites from 1500 to1400 BC. Why did Kenyon fail to recognize the age of the pottery she excavated? She skipped over that and instead focused on exotic pottery that some other dig sites revealed as ancient imports from Cyprus. When she didn't find any Cyprian pottery at Jericho, she reasoned (assumed) that was because Jericho was inhabited some 150 years before 1400 BC. This would have been before Cyprus began exporting pretty pots.

Can you think of another reason why one might not find Cyprian pottery from around 1400 BC in Kenyon's dig trenches? It turns out that her team excavated a poor part of town. Only the nobility could have afforded fancy pots. She should never have expected those Cyprian pots in the first place. She and her team should have instead studied the Late Bronze Age Canaanite pottery that she *did* find. But then she wouldn't have found such fame for supposedly derailing biblical history.

Wood noted at least five other details at Jericho that align with Scripture's chronology.⁶

- Kenyon's excavation reported descriptions of red bricks that fell outward from where they once sat atop a slightly sloped stone retaining wall. Scripture says the walls fell first, then the Israelites sacked the city. If they did not push the walls in from outside, as usual sacks go, then we should find symmetrical piles of wall rubble like we do.
- The fallen mudbrick walls surrounded the whole city—just as expected from Joshua 6. Many of these bricks remain to this day.

- Kenyon even described the walls having fallen *before* the city was burned—again against the trend of burning first and asking questions later. Joshua 6:24 says, "But they burned the city," and this matches the site.
- One section of the wall in the north part of Jericho—nearest the mountains behind the city—did not collapse. One of these intact houses built into that wall would have belonged to Rahab, who saved Hebrew spies according to the Scriptures.
- All excavations so far showed storage jars full of grain. This means Jericho was not plundered, just as God commanded. It also means the siege did not drag on for months like typical sieges, and this matches the one-week siege described in Scripture.

This heap of raw facts from archaeology corroborates too many biblical details for an unbiased historian to ignore. Thus, generally speaking, circumstantial evidence can help reconstruct past events when it piles up, despite its limitations. Kenyon was likely wrong. Her radiocarbon ages were off—not the first time that's happened. The Bible's account of Jericho here remains defensible.

A Key Eyewitness

As helpful as circumstantial evidence can be, eyewitness evidence carries the day in history and law. Maybe this is why the Bible was built largely on reliable eyewitness accounts. Four logical steps through the Bible might add confidence to the picture that emerges from archaeology and that confirms Jericho fell when and how the Bible says it did:

- Hebrews 11:30 says, "By faith the walls of Jericho fell down after they were encircled for seven days."
- We know the book of Hebrews is Scripture because it passes the basic tests of canonicity: it is internally consistent, it treats Old Testament passages as though God wrote them, and it tells the same message as the rest of the Bible—Jesus is God, and He saves sinners who repent and trust Him.
- "All Scripture is given by inspiration of God" (2 Timothy 3:16).

Thus, God inspired the book of Hebrews, which asserts Joshua's history of Jericho.

- And what does the Bible say about the knowledge base of the Holy Spirit who inspired the author of Hebrews? "Known to God from eternity are all His works" (Acts 15:18). God knows everything. He even exists outside of time. Anyone convinced of the Lord's omniscience should thus see that the Lord Himself qualifies as the ultimate reliable eyewitness. Since He inspired both Hebrews and Joshua, His testimony should count.

The unfolding of evidence in and about Jericho was just one example among many of how much weaker empirical science is than what I was taught. Over the years of doing science and sometimes thinking about how it works and how it has failed, I found empirical science not nearly as sure in its statements about origins as it has proven to be in its statements about operations. Armed with this new knowledge of how unsure science can be, I found it easier to process what various scientists were saying.

I found a basic principle that helped discern some fact from some fiction in all kinds of disciplines. If a scientist—including myself—uses past tense verbs, he or she better justify that statement with more than mere experiments. Experiments show how phenomena operate now. They can only offer clues to the past—clues that we tend to shoehorn into our preconceptions. Records tell the most about the past, and experimental results or even just careful observations can align or misalign with those records. I soon wondered how well observations from astronomy matched the record of beginnings that Genesis offers.

Astronomy and Genesis

A Case for Ordered Space

There's nothing like a starry night to prompt us to ponder the meaning of life. I remember one astronomy class field trip to the telescopes. Feeling fairly philosophical, I asked a fellow student what he thought about this vast black and its tiny twinklers. He replied that to him it all looked random. Well, stars are not arranged in some simplistic geometric pattern, but if the matter we see in the night sky were all really randomized, then hydrogens should be spread everywhere instead of bunched into stars.

Since then, I have heard of outer space features that look crafted on purpose despite the many claims about the power of random collisions that experts so often invoke. Collisions made the sun, collisions formed the planets, a collision perfectly positioned the moon for life on Earth to thrive, and collisions crafted galaxies from dust. Collisions don't build anything down here, so I soon started to doubt their power to balance galaxies within their clusters or planets with their moons.

Spiral galaxy ESO 137-001 at 220 million light-years from Earth has young blue stars streaming beneath it.

Who hasn't been told that objects in outer space formed over billions of years? I believed it when that's all I had ever heard, but after I remembered that experimental science is pretty weak when it comes to origins, I began to open my brain to other possibilities. I have heard my astronomer friends describe stellar features that look like they were created recently instead of over billions of years. My college courses never mentioned these features.

We too rarely consider the organization of matter, even in outer space. Conventional cosmologists—who make anti-miracle conjectures about the origins of the universe—keep working within a set of standard models popularly known as Big Bang models. If it all started with a bang, then matter and space should have pretty evenly spread apart. Instead, the universe has more matter in one half than the other half. It also has matter bunched tightly into stars and planets. So what?

Dust doesn't turn into stars, so why does science news constantly repeat the phrase "stellar nursery" as though astronomers routinely witness stars turning on? We do see fascinating space zones where stars illuminate nearby dust. Most astronomers believe stars form in these zones, but none of them has ever documented a star ignition.

Even if we don't see stars form in practice, do the laws of physics at least permit condensing gases to form stars in theory? I have worked with physicists who have run calculations to simulate dust turning into stars. It turns out that the outward push of colliding particles far overpowers gravity's puny inward pull between floating protons or even dust grains. Evolutionary experts know this, so they imagine collisions to supply the extra energy required. Supposedly, shockwaves from a nearby star explosion could compress gas tightly enough for it to start attracting more particles as it evolves into a star. Fine, but what about that preexisting star—the one that exploded? Star origins don't look natural at all. They look decidedly supernatural. In this case, physics fits Genesis 1:15, "And He made the stars also." God—not nature—made stars.

Similar arguments apply to planets. And the discovery of thousands of planets outside our solar system (called extrasolar planets) has baffled classic planet formation theories. The so-called nebular hypothesis calls upon a ring of dust to clump into sun, planets, then moons and comets, all by natural gravity and collisions. This already suffered intractable shortcomings like being unable to explain the planets or moons that spin backwards. The nebular hypothesis fails to account for the unique compositions of Mercury, Venus, Earth, Mars, Jupiter, Saturn, and the rest. If these orbiting bodies came from the same original disk-shaped rotating nebula (and what enormous force supposedly shoved all that material into rotation?), then they should share similar elements, chemicals, and even isotopes. They don't.

Each extrasolar planet looks unique, just as do our solar ones. Hypotheses that featured the formation of solid planets near their suns and gaseous planets farther away vaporized upon the discovery of so many gaseous extrasolar planets that orbit right next to their suns or partner stars. Experts who dismiss the Genesis origins option—who dismiss the supernatural—wanted a one-theory-fits-all origins tale for planets. Too bad.

Why should we find matter organized into solar systems, star clusters, galaxies, galactic clusters, or superclusters as we do? The Genesis concept of a personal God who handcrafted each space object does explain the origin of what telescopes are picking up. Paul picked up that same Genesis concept when he wrote, "There is one glory of the sun, another glory of the moon, and another glory of the stars; for one star differs from another star in glory" (1 Corinthians 15:41).

Wow, You Look So Young

Recently a radio talk show brought me in as a guest commentator. One caller asked a confused question that included random words such as starlight. I couldn't make much sense of the collection of non-sentences I was hearing, so I decided to just name some young-looking features of outer space. After all, no conventional textbook gives them a moment's notice.

Blue stars burn bright. A certain class of blue stars races through its fuel so fast that the stars would die after a million theoretical years. And yet these stars stud the universe. Telescopes spot them both in the cores of galaxies and floating free. They occur near and far, up and down. How can these stars persist in a supposedly 13.8-billion-year-old universe?

Ironically, blue stars may be the biggest reason why naturalistic astronomers believe in stellar nurseries—they need blue star generators so that they can have recently formed stars in a very not-recently formed universe. Some may assume that blue stars formed recently, but until someone observes and reports it, I'm under no obligation to adopt this assumption. While naturalists wait to see a single blue star ignite, why not admit that the existence of billions of blue stars throughout the sky is a reasonable fit with the recent creation described in Genesis?

I didn't have the airtime on that radio show to mention comets. Short-period comets have projected life spans of no more than 10,000 years. Our solar system has thousands of them. Where did such young-looking objects come from? Just like the supposed stellar nurseries, origins-by-nature-only advocates created a comet generator they call the Oord cloud. Nobody has ever seen it. Perhaps the comets look young because they—along with the rest of the universe—actually are young.

The solar system is packed with youthful features, such as moon (like Enceladus) and comet (like Hartley II) geysers that still spout materials into space. Extra heat spills out of planets (like Saturn) and moons (like Io). Visitors to the ICR Discovery Center in Dallas, Texas, learn all about these in our planetarium show on the solar system.

On a much larger scale, evolutionary astronomers face the galaxy wind-up problem. This describes the observation that stars nearer the center of those gorgeous spiral galaxies wind up much faster than the stars near the galaxy's outer edges. When experts apply today's wind-up rate to the past, they end up with galaxies much younger than their billions-of-years expectations. Attempts to solve the wind-up problem so far have invoked gravitational bands to keep stars inside their spiral arms. But these create more problems than they solve. Where did the supposed gravitational bands come from? What has maintained their shapes for so long? Spiral galaxies look relatively young. God may have put the stars in place, then set the galaxy wind-up rate as a special witness of His recent creation to astronomers who too often ignore it.

The very next caller, right after the confused one, asked if we could explain how a recently created Adam and Eve fit into an ancient universe. I wanted to say, "Didn't you hear what I just said? The universe looks young!" but I bit my tongue.

Starlight and Time

But I hear your objection. I have heard it countless times. How can the universe look young if its starlight must be very, very old? To which

I reply, "How do you know that light must be that old?"

I just read the testimony of a Christian-turned-atheist who cited the basic "physics" of starlight as a big reason why he had to abandon the Genesis that his conservative Christian parents taught him. What physics? The argument seems simple on the surface, but it smuggles in some sneaky non-physics assumptions. It goes something like this.

1. Light takes one year to travel 588 million miles. This distance equals one light-year.
2. Distant galaxies are billions of light-years (billions times millions of miles) away.
3. Therefore, the light took billions of years to travel from the star to Earth. The universe must be billions of years old in order for us to see light from so far away.

I see no problem with the logic here. However, the more I read about stars, the more premises I see *missing* from the argument. Add in one or two key premises and the conclusion gets reversed!

The galaxy wind-up problem discussed above confronts this old universe conclusion. As another example, if the universe is billions of years old as this Big Bang-friendly argument demands, then it runs into its own light-travel time conundrum called the horizon problem. The temperature (as inferred from background radiation measurements) of outer space is virtually the same in all directions. The problem comes from the Big Bang's 13.7 billion years, which is not nearly enough time for radiation between distant zones to have crossed paths and evened out their temperatures. If you want to smash Genesis with a light-travel-time club, well, it's a free country. But to be fair, that same club beats against the Big Bang.

Now let's insert some missing premises to see what kind of wrench they throw into the works of the overly simplistic and all-too-popular argument outlined above.

1. Light takes one year to travel 588 million miles. This distance equals one light-year.
2. Distant galaxies are billions of light-years (billions times millions of miles) away.

3. Therefore, the light took billions of years to travel from the star to Earth. The universe must be billions of years old in order for us to see light from so far away.

4. However, the consensus 13.8-billion-year age of the universe leaves *too little time* for light to equalize the temperatures of space in all directions, as they are. (Perhaps a 30-billion-year-old universe would help solve the horizon problem.)

5. Also, the consensus age appears *far too old* to accommodate the youthfulness of distant spiral galaxies. (Perhaps a 500-million-year-old universe would help solve the galaxy wind-up problem.)

6. Therefore, something must be wrong with premise 1, 2, or both.

What could be amiss here? Perhaps premise 1 inadvertently assumes that although light takes one year to travel 588 million miles *today*, that does not necessarily mean that it took that long in the beginning. But now we nudge up to the concept of miraculous origins—a possibility that our secular training drives from our minds but that the Bible asserts in its very first verse. Perhaps the best solution to these varying starlight observations lies in one's willingness to think in supernatural terms.

As a Christ follower, I was willing. I even have one friend who studied these problems as an atheist. One of his coworkers asked him for some help in solving them. Once aware of the problems that plagued his presumed Big Bang, science and logic soon inserted supernatural origins into his thinking. He read the Bible, trusted Christ, and now speaks and writes in favor of both a recent creation and of a loving Savior!

We each travel a unique road. I'm forever grateful that mine led to the Lord as the Maker of heaven, Earth, and of new life in Christ. Meanwhile, creation-friendly physicists have noted that God could have pre-pinned starlight to Earth viewers before stretching out the heavens on creation week Day 4—all in accord with Einstein's well-established relativity theories.[7]

In the end, these options suffer the constraint of extending ways the world works today into the creation week. Life's origins suggest a helpful reset. The most rational (and scientific) way to explain the origin of life is not through the ways that life naturally operates. The laws that govern the way cars work do not describe the way cars were made. Life required a Maker outside creation. Life required a miracle. Perhaps starlight did, too.

Astronomer Danny Faulkner suggests that we should keep the miraculous on the table when we discuss starlight and time. He wrote, "We also can agree that the sudden appearance of plants on Day Three and the sudden appearance of animals on Days Five and Six were miraculous....Then why is it so difficult to accept the possibility that God may have used a miraculous means in order to make those astronomical bodies visible on Day Four?"[8]

Genetics and Genesis

Extreme Organization in DNA

Genetics has much to say about origins, and human origins in particular. I took graduate genetics courses in the mid-1990s while pursuing my master's degree in biotechnology. I remember learning a few things that seemed to support evolutionary origins and other features that supported creation.

One apparently evolution-supporting item that my instructor taught was that only 4% of the human genome coded for proteins. They said back then that this equated to only 4% being useful. Conventional scientists labeled the rest—the vast majority—"junk DNA." Evolution's proponents seized the chance to declare those billions of DNA letters as gibberish left over from eons of random evolutionary processes—all before anyone even did a single test to see if it did or did not have some use in cells. We would call this premature if we're feeling nice and deceitful if we're feeling cynical.

In their view, all this supposed junk supplied a playground for mutations to tinker with. Given sufficient sequences, natural selection of random alterations could generate, given enough time, enough differences to sculpt the first population of people from ape-like ancestors. I remember thinking about how that story sounded pretty defensible.

Then one day our professor reported something he read from outside our textbook. Scientists built an algorithm that detects randomness. They ran some alleged junk DNA sequences through the algorithm to find that this "junk DNA" was decidedly nonrandom! The genetic plot thickened.

Years later in 2005, an international collaboration of hundreds of geneticists who actually did look at DNA function published their first of many results. Under the handy acronym ENCODE, their finds threatened to topple the junk DNA icon. How much of this supposedly useless DNA might be doing something inside cells?

So far, every report from this ongoing collaboration reveals that cells use this mislabeled "junk" DNA all the time. Or at different times. Or in different tissues. But somewhere in the body, during development from an embryo to adulthood and whatever point in between, human DNA is action-packed. Large regions code for RNAs that never turn into proteins. Instead, they latch on to certain sections of DNA to turn it off or on, slow it down, or speed it up.

It turns out that cells use most of their DNA for *regulation*. It's like gas pedals or steering wheels. If you can't regulate how fast or in what direction to drive, then you will soon crash your car, end your trip, and ruin your day. Regulation is key, "junk DNA" was all wrong, and what I once considered a genetic support for evolution turned out to be totally false. Actually, the discovery that almost all our DNA gets used in cells backs the Genesis creation account. God made it. That's why it works so well. Many other genetics discoveries confirm creation, including mutations.

Alive for 900 Years

Remember back in chapter 1 where I promised to say something about Adam's super-long life span? Well, here it goes. Understanding mutations helps us compare the Genesis genealogies with genetics. I had to learn more about mutations than my graduate textbooks taught. They said that mutations arise in DNA primarily from copying errors or substances that damage DNA called mutagens. DNA is made of long strings of chemical bases organized into codes like words in books, only with chemicals in cells instead of ink on paper. Each human body cell contains over three billion bases—roughly equivalent to the information content in something like four bookshelves. Every time a cell divides during an organism's normal growth, a few errors slip through the remarkably efficient and miniaturized marvels that make up DNA replication and repair systems.

Mutations happen in normal body cells, like liver, lung, or lymph cells. If an error in one of those cells garbles the protocols that check and balance cell division, then cancer can result. Basically, one needs only to live long enough and eventually that will happen in one cell or another. But cancer seems not to have happened to Adam—at least, not for over nine centuries. Probably his body's cells were better able to manage mutations, plus his cells had not inherited mutations from hundreds of forebears like ours do today. Was something different about Adam's body and those who lived soon after him? Today, mutations follow from the laws of physics—the law of entropy in particular. Surely those same principles were in place back then. Perhaps early humans' cells were somehow better able to cope with mutations so that they avoided cancer for centuries.

What about mutations that happen in sperm or egg cells? They affect the next generation. This much I had learned in schools. Then I bumped into some technical studies that had not yet made their way into textbooks. I learned that any one of the vast majority of mutations has no immediately noticeable effect on the next generation. These mutations just lurk in a person's genetic background, and then they get stuck there with no way to remove them. Those little copying errors get passed on to all following generations. These nonlethal mutations add up. So, although a single such mutation has no effect, thousands of them begin to erode the effectiveness of cellular processes. Diligent labors of serious scientists have traced several thousand diseases to specific inherited mutations or sets of mutations as the cause.

Then I read genetics studies that measured how fast these mutations happen. In one such study, scientists sequenced the complete DNA (i.e., the nuclear genome) from three generations of a few European families. They found mutation accumulation at a rate of at least 60 bases every generation, but they focused on protein-coding DNA. Since most DNA is regulatory and regulatory DNA must also get copied, it stands to reason that each generation accumulates more than 60 mutations each generation.

That's like copying an enormous encyclopedia every two decades, but each copy picks up almost 100 new spelling or grammar mistakes. With so many thousands of pages, the vast majority of encyclopedia entries would still make sense. But after two or three hundred copies over 4,000 or 6,000 years, you would start to see some mistakes in every encyclopedia entry.

How is this connected to Adam? Well, we know most mutations accumulate relentlessly today, and we have estimates of how fast they do—about 100 per generation. This process equates to a slow-ticking egg timer. A population of organisms can accrue only so many copying errors. Eventually, enough important cellular coding instructions get garbled so that cells no longer work efficiently.

After some more generations, those cells no longer work at all, and if one of those cells happens to be an egg or sperm, then a species has more and more stillbirths. If all the cells in a whole population have corrupted DNA data, then the whole population could die. If all the cells in a whole species have corrupted data in their essential DNA, then the species goes extinct.

We see this happening slowly today. Shrinking population sizes accelerate this march toward extinction. Fewer individuals can mean fewer available DNA variations. For this reason, wildlife biologists try to breed the least-diseased tigers in an effort to slow these big cats' pace of genetic decay. Our grandchildren may be the last generation to see a live tiger or panda.

It also happens with dogs that are so inbred and thus genetically limited that many of their offspring develop genetic diseases or fail to develop at all. Dog breeders cross individuals in a weak breed with more healthy individuals from a different breed. This strengthens the genetic integrity of future dog generations. Endangered wild species may not have such options.

What has all this got to do with Adam and Noah living for hundreds of years?

Let's go back to when genetic egg timers were fully wound up—no mutations had yet entered their DNA. Would we see differences in those ancient DNAs? Of course, we cannot directly measure the health or completeness of ancient DNA, so the best that can be done is to extrapolate what happens now into the distant past. What does it reveal? Our ancestors had fewer mutations than we do. Their cells were healthier. And since these genetic egg timers only count down—no known natural process reverses mutational buildup in sperm and egg cells—we can only go back to when humans had no DNA mutations. Adam and Eve fit that description. They were created with perfect genetics.[9]

Today, DNA repair systems fix the vast majority of those errors. So, with no mutations yet in Adam's DNA repair protein instructions, they would have worked with maximum efficiency. No wonder they could live for so long.

Adam, Methuselah, Noah, and the like lived for nine centuries because they did not inherit hundreds of generations' worth of mutations. Their bodies *must* have worked way better. None of this was in my textbooks.

We can plot the life spans of the Genesis patriarchs. Their life spans did not rise and fall haphazardly like some fantasy might have had it. Instead, they dropped systematically. They curved downward right after the Flood. This makes good genetic sense in light of the dramatic

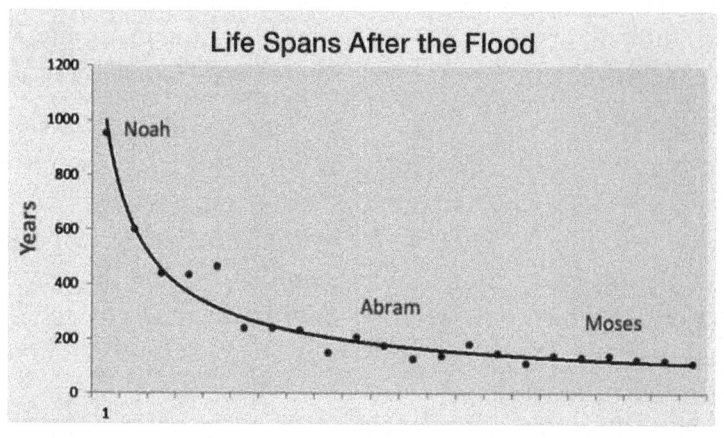

A plot of the life spans given in the Masoretic text of Genesis 11 (found in our English Bibles) shows a systematic trend that should look familiar to biologists. This realism is consistent with the option that early Genesis history is correct.

genetic bottleneck that happened when the entire human race shrunk down to eight individuals.

A 2018 UK news headline read, "Human beings on brink of achieving IMMORTALITY by year 2050, expert reveals."[10] Here is what this expert—Dr. Ian Pearson—bases his bold assertion on: that our descendants living around AD 2050 will be able to replace their organs and that the pace of new health technologies will continue to accelerate. I can't refute either—I don't know the future. But who would want to replace the largest organ in the body—their skin? Sounds torturous, if it could even work. And it would have to work in order to achieve immortality. Pearson, a futurologist, said in the same article, "A long time before we get to fix our bodies and rejuvenate it every time we feel like [for those who can afford it], we'll be able to link our minds to the machine world so well, we'll effectively be living in the cloud."[10]

I don't want to be a Debbie Downer, but just a few short years ago, brain scientists discovered a whole new dimension of neuron connections that instantly multiplied previous estimates of that number of connections. Their new discovery revealed that one human brain has more such connections than the total number of computer and router connections in the whole world. Someone who honestly expects that we can replace our brains must be ignoring brain research. And that's just the brain, not the immaterial aspects of a human like love, longing, or laughter. Each new discovery simply reveals new levels of ignorance about each human organ's inner workings, including the brain. We still have no clue how brains interface with the mind. Pearson may already be living on a cloud.

Back to grim reality, consider the future of this trend. It's only a matter of time. At the relentless rate of about 100 mutations per generation, those who survive lethal mutations will continue to experience the difficulties of accumulated nonlethal mutations. Immune system overreactions, asthma rates, learning disabilities, and body system abnormalities of every stripe and degree continue to climb. Some sociologists are starting to panic over the world's dropping IQ scores. We're getting physically and mentally slower overall as mutations garble the instructions for nerve cell function.

Extinction awaits. This genetic news fulfills God's Genesis promise, "But of the tree of the knowledge of good and evil you shall not eat, for in the day that you eat of it [dying] you shall surely die."[11] The only practical way to thwart this outcome is to reverse every mutation—to restore each individual's DNA to the quality of Adam and Eve's DNA. Who could do that except the One who created perfect DNA in the first place? Mutation accumulation gives fresh appreciation for God's promise to give us new bodies! One of Peter's passages surely takes new bodies into account:

> Blessed be the God and Father of our Lord Jesus Christ, who according to His abundant mercy has begotten us again to a living hope through the resurrection of Jesus Christ from the dead, to an inheritance incorruptible and undefiled and that does not fade away, reserved in heaven for you, who are kept by the power of God through faith for salvation ready to be revealed in the last time. (1 Peter 1:3-5)

Noah's Three Daughters-in-law and Other Finds

Many other genetics discoveries stunningly confirm recent biblical creation. For example, studies of differences in a unique DNA sequence that everyone inherits from their mother's egg cell trace all mankind back to three fundamental types. The research available links these three *haplotypes* to Shem, Ham, and Japheth's wives.[12] After all, "these were the families of the sons of Noah, according to their generations, in their nations; and from these the nations were divided on the earth after the flood."[12]

Further, scientists have measured the mutation rate for that same type of well-studied DNA—mitochondrial DNA, or mtDNA. If today's rates represent past rates, then our mtDNA looks only 10,000 years old—a number not too far from the Bible's 6,000 or so years of human history. The Institute for Creation Research describes these and other Genesis-confirming studies in thousands of free online articles searchable at ICR.org.

Despite the conventional scientists' evolution-soaked opinions about the past, the science of genetics overwhelmingly supports Gen-

Human

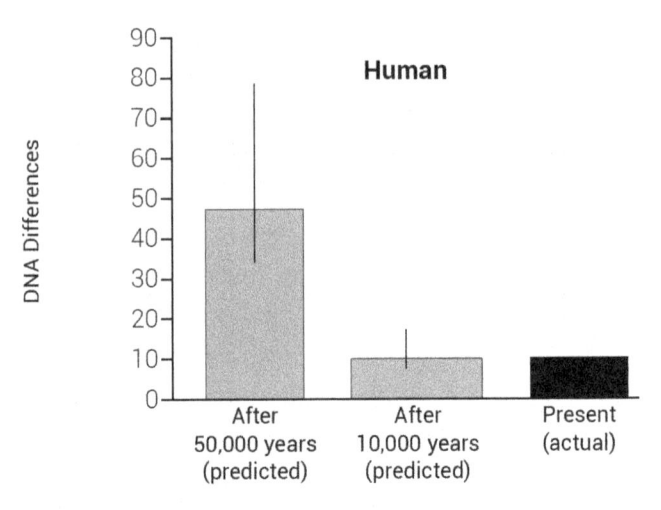

| | After 50,000 years (predicted) | After 10,000 years (predicted) | Present (actual) |

Graph of the recent origin of mitochondrial DNA (mtDNA) diversity in mankind. If humans have lived for the minimum time that evolutionary models assert (50,000 years), then we should have accumulated, based on our measured mutation rate, over 45 DNA differences. However, genetic surveys worldwide show that any two people have about 10 mtDNA differences, the number that only 10,000 years would have produced at this rate.[13]

esis history. The usefulness of almost all our DNA leaves no room for evolution to tinker and looks like an ingenious creation. A relentlessly downward trend of genetic information loss matches the Genesis description of a cursed world. In particular, the sudden drop in life spans after the Flood matches what we know about the severe effects of genetic bottlenecks. DNA traces back to three original mothers to whom our Genesis origins point. Last (for this book at least!), human mitochondrial DNA (inherited from mothers) and Y chromosome DNA (inherited from fathers) have both accumulated only several thousand years' worth of mutations—right in line with human history that Genesis teaches.

Biology and Genesis

Three Biology Basics That Genesis Got Right

Our world relentlessly demonstrates three biology basics that Genesis got right. As God commanded in Genesis 1, each living kind generates more of its own kind and never of another basic kind. Each kind has its variations while sticking to its basic body design. For example, flowers maintain their essential identities even though their petals may display different colors.

Cats exhibit various coat patterns and body sizes, but they remain cats. Even cats with different body and coat types have the ability to interbreed in a continuum. For example, lions can cross with tigers and pumas with ocelots. Genesis 1:24 says, "Then God said, 'Let the earth bring forth the living creature according to its kind.'" Cats, whether they are small or large, whether they look striped, spotted, or smooth, always produce cats.

Similarly, pythons produce pythons. After Hurricane Andrew toppled Miami pet stores in 1992, Burmese pythons (*Python bivittatus*) got loose in the Everglades. Their population has since grown. Scientists recently discovered wild crossbreeds between Burmese and Indian pythons (*Python molurus*). Possibly these hybrids have more genetic gear that will help them pioneer new territory in the southern United States. Maybe they'll slide into your backyard next! Still, pythons remain pythons.

People can breed animals selectively, as Jacob did in Genesis 30, or crosses can happen in the wild. Either way, according to biblical history, the variants, or breeds, would have come from separated populations of the same created kind. Regathering long-separated genes masks the harmful effects of mutations that had been accumulating.

In other words, when evolutionists see a new variation split off from a population, they excitedly supply that new variation a new species name and believe that it may someday evolve onward and upward into a different creature entirely. Odds are the variation merely devolved backwards and downward into a weaker version of the same creature.

Whether plant or animal, living things keep reproducing according to their created kinds. Even the most studied laboratory creatures that scientists have tried to force into changing kinds refuse to comply. Thousands of continuous fruit fly generations and countless generations of the lab bacterium *E. coli* come to mind. They can develop diseases or die in the lab, but if they survive the experiments they merely make more of themselves.

A second biology basic that the Bible got right has to do with the role of death in the lives of animals and man. The Genesis-borne idea of death as an intrusion into a once very good creation pops up all over the place. For example, I remember the first time I saw a famous National Park Service photograph of a monstrous Florida python that had swallowed a huge alligator. This dinner proved too large for the snake body to contain. The gator's bulky carcass ruptured the python from within, leaving both creatures dead—one partly inside the other.

Here's an interesting exercise. Show this picture to someone else and watch their facial expression. See if it doesn't betray an innate sense that this scene does not belong to a happy evolutionary progression toward fitter and better but to a sad intrusion of gore into a world that was supposed to be glorious, like the Garden of Eden. It's just gross! We certainly no longer live in that "very good" paradise God created in the beginning (Genesis 1:31).

Python and alligator died together

The book of Romans confirms the origin of death given in Genesis, saying, "Through one man sin entered the world, and death through sin, and thus death spread to all men, because all sinned" (Romans 5:12). Not just humans, but "the whole creation groans" with pain from sin's deathly result (Romans 8:22). How can we ever escape this place of death? We too deserve the judgment of God. His wrath is revealed against our ungodliness and our suppression of the truth. But praise God that the Lord Jesus took our death penalty and rose from the grave in order to rescue us! After He returns, He promises to build a new world where "there shall be no more death, nor sorrow" (Revelation 21:4).

A third biology basic that Genesis got right ties into its assertion that God—not nature or natural processes like eons of death—made creatures according to separate kinds. We have already noted that, but we need also to note what that implies for evolutionary notions. It means that all members of a created kind should share a set of core attributes. In biology, this takes the form of irreducible structures. You can't alter these structures without killing the creature. This fact fences in creature changes to less critical features, and it fences out molecules-to-man evolution.

We see it in fruit fly studies. Scientists systematically deleted one gene at a time. They ended up with a set of irreducible genes and other critical DNA regions. If genetic engineers cannot alter these without killing the creature, then why should anyone think that natural processes could alter these while still keeping the creature alive?

We see irreducibility in visible structures, too. What should we make of the supposed switch from having a skeleton on the inside like fish or flamingos to having a skeleton on the outside like fleas or fireflies? To transition from one to the other would require rerouting the connection points of every muscle. It would leave any such creature unable to move and thus unable to survive. So, we have a third biological basic that Genesis has suggested. Although certain features like color and size can vary, core features cannot. No wonder they do not change, even in lab experiments that are designed to force them to. Biology shows what Genesis 1–3 describes: kinds make kinds, and death intrudes.

An Engineering Approach to Creature Changes

The way that plants and animals shift or alter their traits lies at the core of Darwinian evolution. Historically, and I suppose even today, some people held a rigid creation perspective. They thought God created creatures with almost no capacity for their traits to change. This gave way in Charles Darwin's day to an opposite swing of the pendulum. Full-on evolution instead treats any and every trait as changeable without any limiting parameters. Darwin's followers now portray a vision of plants and animals having descended from the same population of single-cell organisms long ago. Death of unfit individuals and new trait variations emerging in other individuals supposedly drove paramecia to transform into professors. Does either extreme—unchangeable versus infinitely changeable living forms—match reality?

Well, we just saw that core irreducible traits defy particles-to-people evolution. Creatures maintain fidelity to created kinds even while certain traits do change. Coat color patterns shift on guinea pigs. Different cattle breeds have different horn lengths. But cows stay cows. Perhaps more technically, cows interbreed with bison (to form cattalo and beefalo hybrids) and show signs of fertility with faraway creatures like banteng, yak, and others. We call all these bovines or bovids. They share cloven hooves and that famous four-chambered stomach. So, bovines stay bovines. Examples of variations on a stable theme are legion.

We need a balanced view—a more accurate view—of the way creatures adapt. We might make progress toward such a view by looking at creature adaptations or alterations through an engineer's eyes. But this would require a shift from the all-too-common insistence on, and implicit assumption of, external factors that constantly mold organisms. We tend to treat a collection of conditions as though they constitute a sculptor who models clay creatures. On one hand, we metaphorically say that nature "selects"—like how we say the sun "sets," knowing that physically the earth rotates and the sun only appears to set. But on the other hand (and sometimes in the same sentence!), we have used "selection" as a literal mechanism. How can conditions really select anything since they don't foresee or grasp concepts or parts?

Even some thoughtful evolutionists have expressed their doubts about the effectiveness or even the reality of the imagined concept that nature's conditions function as selectors of the best-adapted mutants to transform traits. One of them wrote that selection "appeared to reify, even deify, natural selection as an agent."[14] Another wrote that "natural selection becomes rather like an occult Power of the pre-scientific age."[15]

If instead plants and animals largely adapt themselves to changes in their surroundings, then we should shift to thinking about internal factors, not mere conditions, as the primary drivers of creature alterations. After all, that's what Gregor Mendel discovered with his famous experiments on pea plants. The plants already had the variation within them—in the form of alleles—to produce smooth or wrinkly seedpods, purple or white flowers, and a few other adjustable traits.

Consider an airplane flying through the air on autopilot as a simple illustration of a group of creatures living through time. The plane approaches a mountain range. Its radar detects peaks in the distance. Internal preprogrammed software interprets the data as a potential threat. Since engineers had already integrated that software with output controls, the preprogrammed autopilot system can signal the wing surfaces and engine power to adjust in appropriate ways at the right time—before the plane smashes into the mountain. Would anyone say that the mountain (a mere condition) was solely responsible in causing the plane to adjust its course? What would have happened to the plane if its radar was turned off?

Now consider snowshoe hares living on autopilot. A hare approaches a mountain range. Its internal thermometers detect upcoming snow and cold. It sheds its brown fur and over the fall season replaces it with white fur just in time to blend in with the snow. Did the snow *cause* the hare to adjust its coat color? Of course not. What about the predators? It sounds simple to Western ears to credit predators with inadvertently selecting snowshoe hares that *can change* coat color according to season. Supposedly the predator eats (or deselects) the snowshoe hares that *could not change* their coat color at the right time.

But what seems simple is not always real. How could mere elimination of individuals from a population perform the expert bioengineering required to tune the hare's system of anticipating climate and deploying a trait variation appropriate to that future change? Where are the endless examples of mere conditions that select just the right biological software and hardware needed to navigate those very conditions?

We do not see conditions engineer anything, let alone creatures. We do not see conditions engineer creatures any more often than we see conditions engineer self-driving cars or U.S. Air Force drones. So, if predators didn't do it, what or who did? Well, if an actual Engineer built snowshoe hares, then we should see evidence of His handiwork. And we do! Scientists in Montana who study hares' hairs noted one year that had consistently cooler temperatures and heavier snow. They wrote, "The completion date of the spring moult occurred 19 days later in 2011, consistent with the month longer snow duration in that year."[16] No predators in sight. No deaths of countless hares over generations. The hares adjusted the timing of their moult to be *in sync* with climate patterns. This reveals internal, engineered climate tracking and response systems.

In hares, as with other living creations, internal detectors monitor conditions. Internal, preprogrammed software interprets those inputs as reasons to change (or not to change). A supernatural Engineer had already integrated hare software with output controls. No wonder it signals its own coat color to shift to white at the right time— before winter snowfall finishes blanketing the countryside.

Why not also, just like the autopiloted airplane, speak of the hare in terms of self-adjustment?

Snowshoe hare going through its color change

This brings the advantage of not having to refer to adaptive designs using vague terms like "emerged," "selected for," or "evolved." Conventional scientists use such phrases as so many rugs under which they can sweep actual trait adjustment mechanisms. Instead, we can find direct correspondence between man-made adaptive designs and God-made adaptive designs in order to clarify causes and their effects. But don't feel surprised if that leads to a renewed appreciation for the ingenuity of the Lord Jesus—the ultimate Bioengineer!

The Institute for Creation Research (ICR) is developing an engineering-based model of creature adaptations called continuous environmental tracking. We notice many examples of plants and animals that continuously track various conditions in their environments. They adjust their own traits to fit and fill those new places in accordance with the Genesis 1 command for creatures to fill the earth, skies, and seas. No death is needed: just detectors, logic centers, and outputs. Perhaps most remarkably, some creatures transmit relevant, tracked data to their own offspring. The baby then develops into an adult with features better-fitted to the conditions its parent detected. Stunning, but real.

Mice babies respond to shock treatments that their parents received. Worms respond to low-calorie conditions of their parents. This is the same with humans, for that matter. Children of starved parents tend to increase calorie storage. In China, for example, many children of those whom Chairman Mao starved in the 1960s deployed this calorie-storage metabolism. Examples of internally driven changes keep piling up.

ICR is running experiments on cavefish. We are testing what conditions sighted Mexican tetra (*Astyanax mexicana*) detect to inform them to produce offspring that resemble the smaller, eyeless, unpigmented cave variety of that same species. We are also exposing a group of the blind cave variety to surface conditions to see what they may be continuously tracking.

Blind cavefish, the Mexican tetra, began increasing its skin pigmentation within hours of exposure to a sunny surface during an experiment at the Institute for Creation Research. This targeted response shows hallmarks of internally programmed adaptability.

As science typically goes, we didn't know what to expect. We thought maybe changes would begin to happen a few generations down the line. We still well may, but meanwhile we were stunned to find that even the adult fish shifted certain traits in mere hours.

Neither our sighted nor blind fish had to die in droves to make way for Darwin's model of adaptation by death of the unfit. No death and no eons were needed for these creature adjustments to happen. We do not yet know the construction of these tetras' sensors, but the rapid adaptations suggest their presence somewhere in these fish.

This matches an ever-growing body of biological evidence for rapid adaptations. Stickleback fish flip fast from large, spiny, and armored ocean forms to small, smooth, and less armored freshwater forms. One group of researchers placed sticklebacks in really cold freshwater lakes. Would they even survive? After three years the sticklebacks had pioneered the place. The researchers had to admit that "marine sticklebacks carry sufficient genetic variation to adapt to changes in temperature over remarkably short time scales."[17] The Genesis creation account suggests exactly where that original genetic variation came from. God put it there to equip these creatures to do what He commanded—fill the earth's waters.

Certain Atlantic snails have increased in size by 22.6% in just the last century.[18] This helps them avoid getting eaten, but it means they have to eat more often. It only took three generations to establish a new finch "species," or stable breeding variety, on the Galapagos Islands.[19] Wrongly dubbed evolution in action, we actually see certain non-core traits shift to better fit various conditions. Speaking of finches, one study even found that the developing bird tissues use math formulas that describe the shape of a cone as they build their beaks inside their little eggs. Tweak one variable in the formula to make it a bit longer and another variable to make the base wider. No death of the unfit. Just designed to adjust.

Models like continuous environmental tracking are poised to replace nature-first models of creature change, and this Genesis-friendly shift from externally driven to internally driven creature changes has the potential to rewrite biology. As this biological revolution unfolds, those who begin to see internally designed adaptive tools will have additional reason to attribute more credit to the Lord Jesus, and less credit to nature, as the real Creator.

Geology and Genesis

Of all the scientific disciplines, geology may be the one that Genesis explains in most detail. What does a bunch of water do when it gets moving? It pulverizes whatever lies in its path. When water slows down, it deposits those tiny pulverized bits into layered sediments. Now imagine an entire globe inundated with water. A worldwide flood should have produced worldwide effects, and it did—in spades.

Rock layers. Every continent has them. They're often portrayed as the fingerprint of vast ages, but recent studies call that into question. Research into waterborne sediments uses large flumes. Sedimentologists experiment with grain sizes and flow rates. Pumps push water along what amounts to one of those lazy rivers we enjoy at water parks, only fine-tuned for science. It turns out that even small grains group together as they roll along in water in a process called flocculation. Each group of tiny grains reaches about the size of one sand grain. Flowing water moves them into layers. Specific flow rates produce specific layering patterns. Fast flow makes graded beds where larger chunks drop to the bottom of each bed. Many rock layers around the world show graded bedding. Often, fossil bones got sorted in this kind of flow. For example, larger dinosaur bones sank toward the bottom of a single bed, and smaller bones or bone fragments settled higher up.

Speed up the flume water a bit and you get cross-beds where dynamic underwater sand dunes stack sediments typically angled at fewer than 25°. Wind forms dunes with sides greater than 25° in most settings. So, wherever we find rock layers with cross-beds at fewer than 25° from the nearest horizontal marker in the rock, we have confidence that water—not wind—put it there. Water-made, cross-bedded rocks occur all over the world. The most important cross-beds I know of occur in a famous Grand Canyon sandstone called the Coconino. Evolutionary geologists have long insisted—and geology texts have long described—the enormous Coconino cross-beds as ancient dry land sand dunes, but they need now to measure the bedding plane angles. Hundreds of such measurements have confirmed a watery origin for the famous Coconino cross-beds.

Mainstream geologists who have insisted the dunes formed in dry deserts have interpreted ancient deserts sandwiched between water-deposited layers above and below. Scoffers then use this wrong interpretation to deny the Flood formed these layers—all of them under water—but an honest look at those pesky angles shows that water deposited even the Coconino. Those cross-bedded layers spread in all directions from their Grand Canyon exposures. Sandstones, shales, mudstones, and limestones—all carried into place by moving water, and most of them extend for hundreds of miles.

Increase the flume speed even more and the beds flatten. Just like graded beds and cross-beds, Earth's layers show plenty of thin beds. Sedimentary rock layers exposed at Grand Canyon show the same basic features seen in flume studies with moving water. Rapid water layers that drape continents match Noah's Flood.

Fossil Proteins and Genesis

Faster Than the Speed of Collagen

In this chapter we have been looking at scientific observations that clearly confirm the Genesis record of origins. These are the results that nobody told me about when I was in college. Had I known them, it might have nudged me to consider the Bible more seriously earlier in life. The topic for this section has all the potential to be a game-changer.

I hope it will have just that effect for a biochemistry major I met at the University of Colorado in Boulder. He asked me what I meant by "dinosaur proteins." I explained that when researchers open up certain bones, they find bone proteins and even whole tissue remnants like blood vessels lurking inside. He immediately admitted that none of his professors had told him anything about this! I showed him a few photographs published in the literature so he could see what I was talking about. Now he has a new, Genesis-friendly fact that he needs to figure out. We hope it becomes a seed of doubt planted in his young faith in the God-denying secularism that pervaded his whole stream of education, from formal instruction to popular movies.

The most common bone protein is collagen. These tiny fibers keep bones together like the straw mixed with pharaoh's mudbricks (Exodus 5:6-7). Collagen strands tether together all those connective tissues that tie bones to tendons, tendons to muscles, and muscles to skin. Collagen is tough, does not dissolve in water, and lasts a long time. Animal skin-turned-leather is over 95% collagen. The ancient world used parchment—much like leather—as a medium for writing, like the famous Dead Sea Scrolls. Jewish scholars from the time of Christ ensconced their library in jars. They hid the parchments in caves where they sat for 2,000 years. Today, the once-white scrolls look brown. The once-sharp edges have frayed. Some have decayed into dust except for small, blotchy remnants that conservators preserve in climate-controlled chambers, like the famous and highly valued Genesis fragment that I once saw on display as I looked past its armed guard. Bone collagen is expected to last even longer than parchment, so though it can last for thousands of years, can it last millions?

Archaeologists began long ago to conduct longevity studies on tough proteins like collagen. Experiments involve subjecting fresh collagen to heat. The increased temperature speeds up protein decay. The same principle underlies our use of refrigerators to delay the decay of whatever loiters in the meat drawer. Researchers track collagen decay at three temperatures. This enables us to generate a plot that relates temperature to the decay rate. College chemistry students learn this as the Arrhenius plot, named after Svante Arrhenius, the Swedish recipient of the 1903 Nobel Prize for Chemistry. Once the experiment is complete, the three decay rates reveal all the variables to estimate age. We can then insert any historical temperature estimate into the Arrhenius equation to discover the decay rate for that particular chemical reaction at that temperature. In the case of bone collagen, the experiments actually include a range of decay reactions including oxidation, deamidation, and hydrolysis that basically turn leather into dust over eons.

These studies project a maximum "life span" for collagen kept at around 50°F of fewer than a million years under ideal conditions.[20] Add bacteria to the bone, and its collagen vanishes fast. Add too much water, and it hydrolyzes. Add radiation, and it breaks down even faster. Why do we care about the speed of collagen decay, and what does it have to do with Genesis?

Over 120 scientific reports in all the major journals and many minor ones describe collagen and other equally short or shorter-lived biochemicals that are still found in fossils. Some even describe whole tissues in fossil bones, teeth, and shells as well as organ remnants like skin, blood vessels, hearts, retinas, and intestines.[21] The fossils bear age assignments of tens to hundreds of millions of years. Now, this is a problem. How can whole tissues like skin and bones, made of biochemicals like collagen, osteocalcin, elastin, keratin, laminin, ovalbumin, DNA, and others—all reported from fossils—persist for that long when decay studies repeatedly show they cannot last but a slim fraction of that time?

Ligaments, Tissues, and Biochemicals—Oh My!

One early sign of fossil bone collagen came from a *Tarbosaurus* in Poland. A 1966 issue of the world's leading science journal *Nature* showed an electron microscope image (a micrograph) of the fibrous dinosaur strands that matched the size and shape of dried collagen fibers. In the following decades, various researchers managed to squeeze reports into different science journals of original biochemical signs from inside other dinosaur bones and various fossils. But the big splash came in a 2005 report in the journal *Science*. It showed whitish connective tissue, red-colored tissue blobs, and red blood cell look-alikes inside tubular blood vessels, all from a *T. rex* femur bone buried supposedly 70 million years ago. That paper grabbed enough of my attention to eventually result in my Ph.D. in the subject. In that effort, I pioneered techniques to detect, visualize, and measure collagen in fossil bones.[22]

Since 2005, hundreds of different researchers have published scores of results showing images, scans, analyses, sequences, X-rays, and you-name-it, often in genuine attempts to show that these tissues and biochemicals cannot really be what they look like. After all, fossil proteins challenge consensus fossil age assignments. Perhaps the fossil protein results represent mistakes or did not come from the fossils but instead from some later contamination. For every author who publishes a paper arguing against the position that these are genuine organics that come from fossils, something like 20 papers describe more proteins and tissues in ever-more samples. Biochemicals in fossils keep piling up, and each one demands an explanation. Removing millions of years from the fossil age assignment solves all the problems. It also fits Genesis history.

A fossil lizard still has keratin protein in its scaly skin. Researchers used a cutting-edge technique to visualize and confirm collagen inside a Jurassic sauropod leg bone from China. Several different independently investigated hadrosaur bones from the western U.S.—plus *Triceratops*, some mosasaurs from Belgium, other extinct marine reptiles from Poland, and *T. rex*—have shown intact bone cells and blood vessels. A handful of these have even passed the gold standard for identifying molecules—protein sequencing. Sensitive instruments called mass spectrometers measure the masses of tiny and degraded protein remnants isolated from fossils. One report found signatures of feather keratin on top of *Archaeopteryx* (an extinct bird) feather impressions but not on non-feather areas.

The most spectacular examples in my mind come from Cambrian and even Precambrian rock layers. Sponges bearing age assignments of over 500 million years still have flexible, intact proteins. Seafloor creatures called beard worms (that look exactly like those still alive today) from these very deeply buried rock layers have still-flexible worm sheaths made of still-flexible protein fibers. Another technical paper described chemical signatures of hagfish slime proteins associated with a fossil hagfish (that looks exactly like living hagfish) from Lebanon.[23]

This brief section represents the tip of an iceberg of evidence. It appears that no continent is immune to soft tissue or biochemical fossil discoveries. They come from every continent but Australia, but their discovery down under may just be a matter of time. Original biomaterials like proteins occur in fossils found throughout the earth's rock layers—upper, lower, and middle sediments. This of course fits the model that Noah's Flood deposited almost all these layers during just one year only thousands of years ago. Each example of a buried biochemical like collagen challenges its millions-of-years age assignment. It's as though all the rock layers were deposited thousands, not millions, of years ago all over the world.

Science and Bias

> *"The stereotype of a fully rational and objective 'scientific method,' with individual scientists as logical (and interchangeable) robots, is self-serving mythology."*[24]
> — Stephen Jay Gould, 1994

At this point, a question may have popped into some of my readers' minds. If sciences like astronomy, genetics, and geology really *support* Genesis, then how could a world full of astronomers, geneticists, and geologists insist instead that the data from their respective disciplines *contradict* Genesis?

I think some of the answer lies in seeing what we want to see or hearing only what we want to hear. It's like the selective hearing that every parent runs into. My wife and I had to switch, every so often, from voice commands to a more tactile translation technique when our little ones pretended like they didn't hear us. Scientists are people, and to a large degree people see what they want or are trained to see. Selective seeing creeps from the recesses of the human heart all the way out to published science papers—thousands of them. One example comes from paleoanthropology, the very subjective discipline of human evolution.

Selective Sight for Homo naledi's *Age Assignment*

Paleo rock star Lee Berger backs his popular missing link that his team named *Homo naledi*. The Perot Museum in Dallas, Texas, which strikes me as a shrine to naturalism, exhibits *naledi*. Selected bits from hundreds of jumbled bones recovered from deep within a South African cave underpin the artistically fleshed-out model. *Naledi* looked like a hobbit with a tiny head.

Its discovery came at a time when every missing link candidate examined over a century had turned out to be a fraud, a man, an extinct ape, or a mixture of human and ape bones positioned as though they belonged together when in fact they never did. It came during a time when new discoveries upended the status of every missing link candidate that seemed like unassailable proof of human evolution just 40 years ago. Berger supplied the new *naledi* option just when human evolution experts again searched for an ape-man candidate that might fit the perpetual void between apes and humans. Perhaps this one would be the first to survive scrutiny as a true transition.

One key needed to validate *naledi*'s transitional status is its age-date. *Homo naledi* must have an assigned age of at least three million years for it to fit the accepted evolutionary timeline for humans' potential ancestor. The process of assigning an age to these fossils ended up offering far too few years. *Naledi* eventually turned up with the wrong age, kicking it out of the list of missing link possibles. But Berger fast-tracked the exposure of *naledi*'s anatomy before the hard work of age-dating had a chance to compile. Unfortunately for objectivity, Berger's group quickly popularized speculations about how *naledi* supposedly fit our ancestry long before a separate team had the chance to sort through its several and contradictory age options.

Selective seeing entered their process of picking an age for *naledi* in the following way. Paul Dirks from Australia's James Cook University led the *naledi* dating study published in *eLife*.[25] Dirks told LSU's Media Center, "The dating of *naledi* was extremely challenging. Eventually, six independent dating methods allowed us to constrain the age of this population of *Homo naledi* to a period known as the late Middle Pleistocene."[26]

Why six methods? If these methods really work, shouldn't any one of them suffice? The need to compare so many different methods on various samples gave the team a healthy pool through which selective sight could sift.

The team applied radiocarbon (C-14), electron-spin resonance (ESR), uranium-thorium decay (U-Th), optically-stimulated luminescence (OSL) in a central age statistical model (CAM), and OSL in a minimal age model (MAM) to *naledi* bones, *naledi* teeth, and cave flowstones that surrounded the fossils.

Independent dating methods should reveal their reliability by giving very similar ages for the same samples and their setting. But these six results landed all over the place. They left an 800,000-year gap between the youngest and oldest results. How would you interpret these data?

Method	Object analyzed	Resulting "age"
U-Th	Flowstone FS1b	~290,000 uranium yrs
OSL (CAM)	Flowstone FS1b	546,000-560,000 fluorescence yrs
OSL (MAM)	Flowstone FS1a	~350,000 fluorescence yrs
OCL (CAM)	Flowstone FS1a	~849,000 fluorescence yrs
ESR	Tooth 1	87,000-104,000 resonance yrs
U-Th	Tooth 1	43,500-46,100 uranium yrs
ESR	Tooth 2	230,000-284,000 resonance yrs
U-Th	Tooth 2	66,200-146,800 uranium yrs
C-14	Bones	33,000-35,000 carbon yrs

Table showing Homo naledi *dating results*[25]

Technical workers call age assignments discordant when they contradict one another. It is extremely common in the lab, and yet nobody talks about it in classrooms. Which one is the real age for *naledi*? Do any of them approximate reality? If so, how can we know? Dirks' team had to come up with something, so they selected an age estimate for *naledi* at 236,000 to 335,000 years. Still, this made *naledi* far too young to be any kind of evolutionary ancestor to humans.

Meanwhile, these results also illustrated how attempts to use natural decay processes like decaying radioactive elements or erosion rates as clocks give discordant results. Very often, researchers can pick from a range of clocks and/or results and ignore those that fail to fit their models. They explain away rejected results. That's not science. It's selective seeing—also known as cherry-picking.

The *naledi* dating team chose the older tooth results from ESR and the U-Th ages from certain flowstones. They did not explain why they rejected the U-Th result from three different *Homo naledi* teeth, which ranged from 43,500 to 146,000 uranium years.

The team also rejected all the data (carbon dates) taken directly from the actual bones they were trying to date. At least they offered an excuse this time. They speculated that calcite precipitation made the carbon ages look much younger than the actual (selected) age range. Did they find any direct evidence of contamination in the samples? No—they just reasoned in a circle from the presumption that their selected age was right.

For that matter, if calcite precipitation or any other of a number of isotope-altering processes can explain away this carbon age, how do we know when they should not also explain away other carbon ages? In other words, this "explain it away" option invalidates the whole enterprise.

They have no objective way to select from their options. Perhaps none come anywhere close to the fossil's actual age, yet the team selected some of the oldest "age" results. Even that age landed nowhere near the three million years needed to place *naledi* in the still-vacant missing link spot on the imaginary human evolution timeline. In the end, this exercise typifies the selective seeing that runs rampant through the hallowed halls of conventional science.

The Heart Behind the Bias

Evidence from dozens of disciplines points to a recent creation and a recently flooded world. Why do most scientists miss it? Many have a bias against the very possibility. For them, even as it was for me when I began reinvestigating origins, to merely allow the door toward Genesis to crack open could expose us to ridicule from friends, family members, and colleagues. To consider, let alone embrace, Genesis as history threatens careers, friendships, and the all-important esteem of others. Think of the naturalistic thinker who has built a career on evolutionary ideas. Would such a person have enough bravery to admit they were wrong?

Fortunately for me, I reached this point in my journey of discovering science that confirmed Genesis when I was only 20 years old. I had of course begun my journey with complete confidence in the fact of evolution, but I had not yet built a career upon naturalistic deductions. Determined to find truth even if it would rob me from being in the cool crowd, I decided that sticking with Genesis, if it was true, was worth suffering any disdain my family, friends, or employers might toss my way.

Perhaps at this point some of my readers might identify with Joseph of Arimathea, a wealthy Jewish leader during the life of Christ. This Joseph belonged to a group so tied into their own religious beliefs that they refused the very possibility that Jesus could have been God, despite the proofs of deity they beheld. They selectively saw just those Scriptures that predicted a Messiah who would conquer Israel's enemies but skirted those Scriptures that predicted a suffering Messiah, like Genesis 3:15, Psalm 22, and Isaiah 53.

When Jesus came to live among us, He provided evidence that He was the real deal. And talk about unmistakable evidence! The Lord healed the lame, commanded the weather, and raised people from the dead right in front of those staunch religionists. The overwhelming validations of this suffering Jesus as the Messiah made no difference to them—well, to most of them. Joseph of Arimathea and doubtless a handful of other brave souls followed the evidence where it led—to Jesus as Maker and Messiah.

In the same way, evidence from within our own souls that I discuss in a later chapter plus evidence from science discussed just now may have no persuasive effect on most Genesis disdainers, but the evidence eventually got to me. It also made a difference in the hearts of my colleagues at the Institute for Creation Research and thousands of others who, like odd tunas who refuse to school with their oceangoing, God-denying cohorts, trust and even come to defend the God of creation and His book of beginnings.

The evidence Jesus brought to bear concerning His true identity as God may not have dented most Pharisees' hearts, but it looks like it penetrated that of Joseph of Arimathea. Luke said about him:

> Now behold, there was a man named Joseph, a council member, a good and just man. He had not consented to their decision and deed [to crucify Christ]. He was from Arimathea, a city of the Jews, who himself was also waiting for the kingdom of God. This man went to Pilate and asked for the body of Jesus. (Luke 23:50-52)

By the way, Isaiah 53:9 foretold Joseph's role of supplying a rich man's tomb for the suffering Servant five centuries before he fulfilled it. But the point here is that despite so many of his peers who ignored overwhelming evidence that Jesus was the Christ, this man accepted the evidence for Jesus and may well also have accepted Jesus as his own necessary and sufficient Savior. Despite my many peers who flatly deny Genesis, I now accept both the book and its divine Author. He made sure His representatives wrote the stories of the past even as He now writes my story of redemption and return to truth.

Everything began to change for me once I learned that science confirmed Genesis. I began to retool—or more accurately, I began to let the Lord retool—the way I thought about everything else in this life and the next. If astronomy doesn't really have the answers that it claims to give about the origin of the cosmos, then perhaps God's Word on the matter deserves more attention. If genetics shows a recent origin of mankind through a buildup of only thousands of years' worth of DNA differences between any two people, then I may actually find my origins in Adam, not apes. And that means I belong to God, not to nature or myself!

Now that I see creatures adapting to their surroundings by design, I am compelled to cease attributing clever design to natural processes and instead extol the Designer of life for His fantastic and fearful handiwork. Lastly, the basics of geology include vast sheets of water-deposited sediments studded with fossils and their proteins. Since these point to Noah's Flood as real history, I felt justified in taking seriously God's warnings to accept His offer of salvation from my sins so I can escape any future judgment. Until judgment day arrives, I get to live for Him! I now know God is fully worthy of all my devotion. Once I saw how firmly the sciences support Scripture, He became my source, inspiration, and goal.

I then began to reinterpret all realms of human life no longer from the standpoint of man as the center of all things but from the standpoint of God as the center of all things. And not just any God—the God who cares so much about each one of us that He left us a defensible, believable record of where we came from. What would my world look like if I believed what He said in Genesis? What would a whole society look like if its people held the Word of God, including Genesis, in high regard? For that matter, how might all of history's attempts at manufacturing happy societies tie back to our origins in a happy garden in the beginning? These are the subjects of the next chapter.

4
CULTURE WITHOUT GENESIS

> *"People who never heard of Strauss, of Bauer, or of Tübingen, are quite prepared to say that our Saviour was but a well-meaning man, who had a great many faults, and made a great many mistakes; that his miracles, as recorded in the New Testament, were in part imaginary, and in part accountable by natural theories; that the raising of Lazarus never occurred, since the Gospel of John is a forgery from first to last; that the atonement is a doctrine to be scouted as bloody and unrighteous; that Paul was a fanatic who wrote unthinkingly, and that much of what bears his name was never written by him at all. Thus is the Bible rubbed through the tribulum of criticism from Genesis to Revelation, until, in the faith of the age in which we live, as represented by its so-called leaders, there are but a few inspired fragments here and there remaining."*[1]
> — *Charles Spurgeon, 1886*

What People on the Street Believe About Genesis

I have made three basic observations so far in this book. First, Genesis asserts that creation happened by supernatural, not by natural, events. Genesis even specifies the order and timing of those miracles and the identity of the miracle Maker as God. Second, though they suffer widespread ridicule for their supposedly senseless assertions, Genesis beginnings do explain characteristics of real life and relationships that everybody knows, like death, justice, and grace. Third, the sciences support Genesis beginnings. I had heard so often that Genesis was crazy talk, but once I actually read it and compared it with real life and with the real world, I found in it a more respectable record than I had been told.

But how respectable is that record? Should I take it all at face value? Does it really convey the truth about origins, untarnished by the mud that so many authorities have slung upon it? One approach to answering these questions involves consideration of how attitudes toward Genesis impact cultures. Is a life without Genesis livable? Does a whole society thrive, survive, or fade when it carries a low view of Genesis? If we can personally or corporately live well in the absence of respect for Genesis as history, then I think that counts as a point against Genesis. And vice versa.

Certainly, life in the absence of Genesis history is livable in the sense that one can reject Genesis entirely and still find food, shelter, and clothing. Folks can go to church, call themselves Christians, agree with statements like "Jesus died for my sins," and all the while shake their heads at Genesis. I have lived with them. I was one of them. And I was confused.

Man-on-the-street experiences like those recorded in the free online documentary *Evolution vs. God Uncensored* help explore the question of what folks believe about Genesis—the first step to evaluating the impact of regard for Genesis on culture.[2] It featured atheistic professors and evolution-believing students on a university campus in California. Anecdotes like these may not meet the statistical standards that polls can pull in, but they are real people telling us what they really believe. And they show that most students value what they hear from teachers and scientists as more authoritative and trustworthy than their own vague notions about Genesis.

The Faith of Our Culture

The video *Evolution vs. God Uncensored* opens with a quote from the world's most outspoken atheist, Richard Dawkins. Dawkins said, "Faith is the great cop-out, the great excuse to evade the need to think and evaluate evidence."[2] At least he reveals what he means by "faith." To atheists like Dawkins, it means choosing to believe in something despite reasonable evidence against it. Like choosing to believe that life exists on Mars despite the evidence from probes that the Red Planet is utterly hostile to life.

Many conversations have taught me that too many churchgoers have adopted this caricature of faith. Rather than step through the details of sciencey-sounding claims that refute Genesis, we too often sidestep the whole enchilada with a platitude about how faith needs no evidence—I just choose to believe it. Yuck! Not only does this "evidence doesn't matter" attitude ignore one of the main thrusts of the New Testament—to reveal the historical evidence for the resurrection of Jesus—but it earns deserved disrespect from a world full of skeptics who might consider Christ if a Christ-follower would present a reasonable case.

Acts 17:14 says, "And some of them were persuaded; and a great multitude of the devout Greeks, and not a few of the leading women, joined Paul and Silas." Persuaded by what? By the evidence: the evidence of Paul's life and of his words; the evidence of God's Spirit in Paul and others whose supernatural boldness enabled them to face persecution; the historical evidence that we still teach today for the lordship and resurrection of Jesus Christ, not to mention the evidence from within the Bible that it is divinely inspired, including its internal consistency and fulfilled prophecy.

Faith is not a leap into the dark against all reason. Dawkins got this one just as wrong as his argument that eyeballs are wired backwards—proven false by the later discovery that the biophysics of the human visual system appear to have been "optimized"—a sciencey word for *cannot be improved upon*.[3,4] Rather, faith is a step into the light of reason, revelation, and good science.

"Now faith…is the evidence of things not seen" (Hebrews 11:1). We can't see physical coins or bills sitting in our bank accounts, but every time we swipe our debit cards we exercise a kind of faith that the money does reside there. Dawkins wasn't there to witness his supposed evolutionary ascent from ape-like ancestors. Neither was I. But Genesis asserts, and other Bible passages agree, that someone was there from the beginning. According to Scripture, this Someone knows everything and conveyed through the ages and preserved for this very generation a written record of creation, not evolution, through His prophet's pens. Should we believe it?

Our culture seems to take its cues on faith from atheistic philosophy. Even Christians perhaps unconsciously have redefined faith in a way that ends up accommodating evolution's nature-only ideas. For too many churchgoers, faith no longer means "taking God at His Word—whatever He says." If it did, then we could no longer take scientists at their word when the history they convey so clearly clashes with the Genesis record. So, instead of investigating the word of the scientists or the Word of the Creator, we redefine faith to mean "believing in God despite what science says"—and by *science* we really mean *evolution*.

Scripture calls Christ-followers to a higher standard of belief than what God-deniers demand. The decision to take as true mere portions of God's Word must have some value to God, but He surely esteems an attitude of total trust. Is it any different with us as parents? I have said something like, "Trust me, my dear daughter. If you study for an hour a day for a week, you'll get a better grade on that test than if you study for seven straight hours the night before that test." Sometimes we get to rejoice together when our kids take us parents at our word, follow what we say, and reap the benefits of that wisdom! I see no reason why our heavenly Father would rejoice any less when we do the same with His Word.

Verses like the following began to challenge the weakness of my faith. "If you will not believe, surely you shall not be established" (Isaiah 7:9). How much should I believe? "The entirety of Your word is truth, and every one of Your righteous judgments endures forever" (Psalm 119:160). Should I really believe it all? "I will worship toward Your holy temple, and praise Your name for Your lovingkindness and Your truth; For You have magnified Your word above all Your name" (Psalm 138:2). Do I hold God's Word above the honor due His very name? Our culture generally distrusts Genesis and the Bible, and even Christian subcultures distrust some of Genesis. Few indeed agree with the psalmist's attitudes toward God's Word. How does this frail faith impact our lives and our Western societies?

Mixed Beliefs

Some time back, I had an opportunity to summarize in 30 minutes my five-year-long creation conversion story for a small audience at the Institute for Creation Research Discovery Center in Dallas. A friend of mine brought an atheist along. I respect atheists when they attempt to live a consistent worldview. My friend, who also believes the Bible, relayed to me the atheist's main conclusion about my presentation. She said, "He thinks we're crazy." Truly today, to trust Genesis as history is to be called crazy. Does this kind of peer pressure influence would-be believers to take a low view of Genesis?

I have become convinced that many Americans who identify as Christians would describe their view of Genesis like this: it contains some truth but does not qualify as accurate history. Of course, the range of views on Genesis spans from those who outright scoff at it to those who believe that God Himself was responsible for both inspiring and preserving every word of it through the millennia until this day. Most churchgoers probably fall somewhere in between. How defensible are these in-between views?

Results from a 2013 Pew Research poll outlined American attitudes toward Genesis.[5] About a quarter lean toward atheistic evolution. This regards Genesis as total myth. Others identified more closely with theistic evolution, which supposes God guided the transformation of, for example, worms to wombats over millions of years. Some respondents loosely aligned with biblical creation. This position holds a high view of Genesis by treating it as straightforward history. Most respondents mixed answers that would characterize different views, revealing plenty of confusing overlaps. The big news at the time was the decline—for the first time in decades—of the creation view. As the Baby Boomer generation gives way to the emerging Millennials' kids, fewer folks respect the Genesis account.

Another poll confirmed this rising tide toward evolutionary ideas. In 2012, Gallup found that about 46% of Americans held to a general Genesis account of origins.[6] The number hovered around there for a few decades, but a 2017 Gallup poll asked the same set of questions to find that number suddenly dropped to only 38%.[7] The atheists' messages seem to be making their mark, and fast.

Nowadays, nobody who's anybody thinks Genesis is cool. Now, if we all arrived through creation according to Genesis, as I have become convinced, and not through evolution according to popular culture, then we are in serious spiritual danger. That's why I wrote this book. I want to present a case for getting back to believing in Genesis and in the God who brought this written record through Earth's millennia all the way to our eyes today.

What do folks on the street believe about Genesis? Today's street-level philosophy assumes that those who really believe in Genesis are *backwards*. But if Genesis is right after all, then we're getting our *backwards* backwards. Thus, much of this book concerns itself with naming those babies we end up throwing out with the apparent bathwater of Genesis. We want those babies, but can we really separate them from Genesis origins? On the other hand, is Genesis really as indefensible as most folks seem to take for granted? I had to find answers to questions like these. So, I looked to cultural influencers. What do those who, for example, train our pastors teach about Genesis? And what evidences or influences led them to their positions?

What Christian Intellectuals Believe About Genesis

"You will understand that my rationalism was inevitably based on what I believed to be the findings of the sciences, and those findings, not being a scientist, I had to take on trust—in fact, on authority."[8]
— C. S. Lewis, 1955

Does it matter what Christian scholars believe about Genesis? It makes a huge impact. These valuable thinkers write the big, thick academic books that turn over every leaf and examine each facet of an issue. They set an influential tone. They teach teachers. They train preachers who influence millions, and they directly inform more masses who read their works. And like all of us, to one degree or another, preachers parrot what they learn from their authorities. Unfortunately, too few Christian scholars tackle the challenges that Genesis appears to present. Their silence is remarkable. Here we have *the* book of beginnings—to which all Christian doctrines reach back—and even those scholars who love the doctrines shy from the creation, Flood, and Babel accounts in Genesis. Why?

Pastors have approached me several times after my creation presentations. They tell me that they never knew how well science supported Genesis. They say that their seminary never touched it. Zero classes on creation. No in-depth studies of early Genesis. Why not?

Two Histories

The primary interpretive challenges with Genesis involve conflicting histories. Genesis conveys one history of ultimate origins (summarized in chapter 1 of this book), but the world's naturalistic scientists convey another, very different history. To confuse matters, these scientists call their version of history "science" despite the fact that empirical science deals exclusively with repeatable observations, exactly unlike history.

The conventional origins story goes something like this: Billions of years ago, all matter, energy, and even the space in the universe suddenly exploded outward in a Big Bang. Almost immediately afterward, the universe's expansion rate radically inflated. It later slowed down. Then, matter began condensing and organizing itself into stars and galaxies. Generations of stars self-ignited and fizzled out. When they exploded, they made space dust that clumped together to form planets near stars. Countless collisions of some of that stardust formed a hot Earth and moon.

Water later arrived (or perhaps it was just there at the first) to eventually help spontaneously generate the first population of living cells. As soon as the cells began copying genetic material and making mistakes in the process, they could generate subtle differences between them. Over billions of years, various natural factors like heat, salt, air, wind, and predators killed some individuals while others survived and continued to change. This process eventually produced all living things, including mankind, from those first cells, which came from stardust, which came from a quantum fluctuation of nothing that exploded into everything.

Some newer versions invoke many universes—the multiverse theory. At least that way the improbability of our expertly designed universe gets drowned in an infinite sea of endless probabilities—for which no evidence exists.

And they call this science. It's not, but too many seminarians seem unfamiliar with the difference. This grand evolutionary narrative boils down to a man-made story into which various facts are made to fit. Can the same facts fit Genesis history? Absolutely—and with far fewer fatal flaws.

Science is supposed to describe observable events. Who observed the origin of the universe or the first cell supposedly emerging from matter? Certain characteristics of starlight and the space that surrounds it suggest an expanding universe. Even if we call this expansion an observation, that doesn't mean that events described in a Big Bang model caused it. Possibly God created the world at a certain size and set it up to expand from there. Similarly, God, not nature, created cells. Ah, but then conventional evolutionists play the anti-God card, saying that science can test no hypothesis that includes God (who by definition resides outside of nature). In saying this, they overlook the fact that their story of evolution also lies outside of repeatable natural phenomena. In other words, it comprises historical statements that use past-tense verbs and thus lies beyond that which we can directly measure.

The popular conventional approach also ignores the possibility that God is the best explanation for certain observations. For example, what about the unnaturally precise design that creation shows from the largest to the tiniest of scales? At a large scale, the Earth-sun-moon system holds in precise balance mass, energy, distance, and trajectory, all to sustain life. At a nanoscopic scale, DNA encodes—literally—a unique language. What scientist has ever recorded raw nature crafting such specifications? These observations should reel a real Creator back into the bag of possibilities.

Origins Stories

Bottom line, naturalistic scientists tell stories but declare them as science. All the while, their speculations reflect more human imagination than any experimental results. Worse, these imaginings intentionally circumnavigate God even in the face of undeniable evidence for His miracles of creation and the Flood! Paul describes such creation-scoffers in Romans, saying, "For since the creation of the world His invisible attributes are clearly seen, being understood by the things that are made, even His eternal power and Godhead, so that they are without excuse" (Romans 1:20). And since theologians don't receive training on how to discern between scientific and historical studies, they feel intimidated into adjusting Scripture to accommodate whatever the scientists say happened, or they feel pressured into silence on the whole matter.

Of the relatively few scholars who do write about Genesis, most suggest it conveys perhaps bits of history but is mixed with myth to convey spiritual lessons. To many, Adam and Eve weren't real people but literary tools to illustrate something about God—perhaps His holiness that "they" violated. To these authors, God did not create "the heavens, the earth, and all that is in them" in six days (Exodus 20:11). Rather, for them the Genesis creation spiel illustrates something about how a God has some power.

Likewise, Noah's Flood never covered the globe. After all, it would be scientifically impossible to cover the Rockies and Himalayas with water—and where did all that water go? According to this worldly reasoning, the best the Flood tale can tell us is that God's grace may somehow reach toward mankind. While we're here, suffice it to say for now that Genesis informs us the Flood's waters covered "all the high hills" of the pre-Flood world (Genesis 7:19), not today's massive mountain ranges.

The late theologian John Walvoord summarized the general situation well. He said that key 19th-century German theologians taught that Bible readers should focus on Scripture's spiritual intent only. We should divorce its heavenly things from its earthly things like history.

> This led to contemporary liberalism of the twentieth century which assumes that the Bible cannot be taken seriously in its historical or factual content, but considers Scripture [especially Genesis] only as a means of gaining spiritual insights.[9]

But why should anyone trust spiritual insight from a document laden or at least sprinkled with errors?

Some of my friends have never heard this resistance to Scripture. They grew up simply believing the Bible, including all its history. They express shock when I express the apparent faithlessness among many Christian scholars. They ask, "How could we consider them Christians if they don't even believe the Bible?" Well, they have trusted the Lord Jesus for salvation from sins. Many professing Christians believe most or some of the Bible even though they *say* they believe the entire Bible. I have come to believe that they—as I and the Pharisees once did—replace the *words* of Scripture with their interpretation *of* them when it is convenient. Thus, some assert that God never intended the "unscientific" parts (Genesis 1–11) to convey history. If He did, then "science" wouldn't have refuted it.

This brings me to the very purpose of the ICR ministry. We exist to inform the church or whoever else is interested that Genesis 1–11 does make scientific sense after all. The key difference comes when we discover, as I did, that evolution and its millions of years do not qualify as science. They boil down to speculations about the past made by men and women who were not there. They never saw the supposed events that they assert with such confidence, and the evidence we all can see has a better-fitting, creation-friendly explanation.

Linda's Return to Genesis

A friend named Linda endured a trying spiritual journey that illustrates what intellectuals believe about Genesis as well as the importance of Genesis to Christianity. She grew up as a rural, west Tennessee preacher's daughter who followed her parents' views. They taught her that the Bible was totally true. Unfortunately, they never prepared her for the onslaught of Genesis doubters that she would encounter among the high-Christian-character professors at her Christian college.

She took an Old Testament survey class from a generous, kind, godly old professor who attended church every Sunday and supported missions. He taught them that Genesis does not hold history. Linda heard that the book cannot be believed as written. Her professor asserted that Hebrew scribes concocted Genesis thousands of years after the fake events that it portrays. She was devastated. Her confidence in the Bible as the Word of God collapsed. She dropped out of school after that. She stopped going to church. She told me years later her thoughts: if Genesis was fake, then there was no telling what other parts of Scripture were also forgeries. Her life spiraled far afield as she explored other belief systems. None of them brought satisfaction, and after years of wandering both geographically and spiritually, she was no nearer the answers to three of mankind's most fundamental curiosities:

1. Where did I come from?
2. Why am I here?
3. Where am I going?

Linda found herself in Switzerland one evening. She was watching the sunset in the Swiss Alps, and it was stunningly gorgeous: orange, purple, red, and yellow mixed with white clouds, blue skies, and striking snow-capped mountain peaks. It looked better than any man-made painting. The sheer splendor struck her so pointedly that she suddenly knew nothing so lovely could happen by chance. There had to be a God. So, she prayed for the first time in many years. She said simply, "God, if you are real, show Yourself to me." And He did! A few months later, she came across one of the early books written by ICR founder Dr. Henry Morris.

He and theologian John Whitcomb coauthored the seminal 1961 book *The Genesis Flood*. It detailed, to the satisfaction of many Christian intellectuals, how the Flood explains the earth's surface as we see it. It described experimental and observable evidence that large volumes of water can erode canyons and deposit layers fast. It suggested that the very reason why most fossils exist is because of their rapid burial in the watery catastrophe of Noah's Flood. Its arguments largely remain valid today, even though the authors wrote the book before plate tectonics became established in Earth science. By absolving the need for millions of years to deposit rock layers (we just need lots of water for that), multitudes found renewed confidence that the Bible got it right all along. That's what happened to Linda. Her eyes were opened to the defensibility of the idea that Genesis is history and not allegory. She discovered at last that her Christian professor, who had wrecked her faith, was simply wrong.

Theologians' Confusions

I hear similar stories over and over as I travel and speak on creation issues. Today's generation of Christian scholars seems to have adopted its views on Genesis from their intellectual forebears. One of my co-workers recently took an online course on Genesis from a prominent online American Christian university. He was very impressed by the deep and meaningful insights in the required readings from their Genesis commentary textbook, but he was equally shocked to discover that the same insightful and obviously brilliant author suddenly switched to a different protocol when he interpreted Genesis 1. At that point, the author abandoned the idea that the context of the text should frame its own words' meanings. Because scientists insist that creation could not possibly have happened the way Genesis says it did, this author concluded that Genesis 1 was never meant to convey history—just certain theological themes. How inconsistent of him to treat the same style of writing in two totally different ways. And why? All to accommodate stories that Bible-denying scientists invented.

But what clues did God leave in His Word that suggest we should treat some of what sounds like history as history and other similar phrases as not history? He never tells within the text when to switch from one to the other. Does He even want us to know where we came from? Of course He does. But that means we need to treat His history as historical. The Lord Jesus put it this way to the Jews who were persecuting Him:

> "Do not think that I shall accuse you to the Father; there is one who accuses you—Moses, in whom you trust. For if you believed Moses, you would believe Me; for he wrote about Me." (John 5:45-46)

These religious Jews had memorized Moses' books and thus *knew* the words, but they didn't *believe* them. Similarly, He has told us Christians through Moses' writings about earthly things—the creation of the earth, the curse upon the earth, and the Flood across the whole earth.

When we believe the words of Moses, including Genesis, are historical—which good science supports—then we more readily believe in Jesus as the Savior according to that same source.

When it comes to early Genesis, does the Bible give Christians latitude to abandon the usual way to understand the Bible's text? Did the New Testament authors use different interpretive methods when they quoted Genesis or discussed the implications of various Old Testament Scriptures? Have Christian intellectuals esteemed the pronouncements of fallible humans (scientists though they may be) over the pronouncements of an infallible God when they insist that Genesis cannot mean what it appears to mean? A key purpose of this book is to tell the patient reader how I, for one, came to believe that God made no mistakes in Genesis. I aim to show why every Christian should feel confident about the history in Genesis despite what certain intellectuals say.

What Bible Believers Say About Genesis

Despite the avalanche of disdain aimed at Genesis in today's Western societies, men and women from all walks and realms still trust the ancient words of Genesis as accurate. We take the Bible's words to mean what they plainly, though not always literally, say. For example, when the text in Genesis 1 uses the term *day*, we let the context define its own meaning rather than forcing an outside meaning onto the word. It defines *day* as that time which spans from evening to morning. It leaves the unbiased reader with no doubt that those six days of creation were normal, regular days like we have today.

Many Christians who force millions of years into a history based on Genesis accuse those who take the plain meaning of its words of being wooden literalists. Some of them argue that since Genesis was written in the form of a poem, its words should not be understood plainly but symbolically. Now, this principle does apply to parts of the Bible but not to the historical parts.

Historical narrative uses past tense verbs to relay events in sequence. For example, "I went to town last week to meet a friend and found a newly opened store. So, I stopped in to see what they had and decided to buy this shirt." In contrast, poetry uses various verb tenses in structures not intended to convey events but to convey concepts, feelings, or both. Consider Psalm 31:3, which says in part, "For You are my rock and my fortress." God the Father is spirit, not a literal rock or fortress. Here, David the psalmist uses these words as simple metaphors that express the safety that God supplies. The Lord is like a rock or a fortress in that He protects. Rocks and fortresses are hard for David's enemies to get through. God would be tough to get through, too.

Further, whereas today's poems tend to include rhyming or alliteration, Hebrew poetry tends to stack parallel statements in its poems. For example, the very sad Psalm 88 has two parallel statements in verse 5:

> I am set apart with the dead,
>   like the slain who lie in the grave,
> whom you remember no more,
>   who are cut off from your care.

These poetic utterances differ entirely from the blow-by-blow relay of events found in Genesis, Exodus, Kings, and other history books. In ignoring the historical genre of Genesis, those who argue for its treatment as not history relinquish the means to understand what it says. This opens the door for any Scripture to mean anything and thus closes the door for God to say anything meaningful.

Taking the Genesis chronologies as history leads to a 6,000-or-so-year-old world. It means believing in its two biggest miracles—the creation and the Flood. "For in six days the LORD made the heavens and the earth, the sea, and all that is in them, and rested the seventh day" (Exodus 20:11). And "the world that then existed perished, being flooded with water" (2 Peter 3:6). How should Bible believers treat Genesis? Like it says exactly what it means on its surface. God spoke the everlasting truth about our past and did so in a way that folks from any culture could understand it.

What I'm saying is that too many Christians in our Western nations don't believe in their own Bibles. We have been intimidated by so-called science until we retreat into corners. No wonder we have so little impact on our societies. And no wonder we had too little an impact on societies of the not-too-distant past. Let's look at a few examples before we consider more benefits of returning to Genesis as history.

Communism's Position

We can learn helpful lessons from large-scale crimes against humanity. What happens when societies try to function as though we did not come from "Adam, the son of God?"[10] Take the case of the nefarious Joseph Stalin. He led the U.S.S.R. to stomp out or starve more people across Eurasia than perhaps died fighting in World War II. The death and pain he spearheaded under the banner of communism compares with that of his communist cohort in China, Chairman Mao. Both viewed people as creatures with about the same inherent worth as worms. Tens of millions died at their hands. Communism faded from former U.S.S.R. lands, but atheistic philosophy still reigns in Russia, China, and other nations. Communism still rules China, where the populace have few rights, and those are considered gifts from the state instead of inalienable from the people. More humans died in the 20th century than in all wars in all history prior—and most at the hands of communist regimes.

What is communism, why has it proved so destructive, and what do its devotees believe about Genesis? Communism has roots in the philosophy of Karl Marx. Like his contemporary Charles Darwin, Marx viewed the world as a big struggle for survival.[11] The complete title of Darwin's most famous work, *On the Origin of Species by Means of Natural Selection, or the Preservation of Favoured Races in the Struggle for Life*, published in 1859, expresses the core concept of struggle that Marx borrowed and that millions of communists incorporated into human government. Marx's friend and coauthor Friedrich Engels wrote, "Just as Darwin discovered the law of evolution in organic nature, so Marx discovered the law of evolution in human history."[12] Of course, these "laws" existed only in their minds, not in reality. On June 16, 1873, Marx inscribed a note to Darwin on the inside of his newly published English language version of *Capital* that said it was from a "sincere admirer."[13]

What did Darwin do for the sciences? His concept of natural selection swept the Creator out of the sciences. Who needs God to design life forms when nature does such a fine job all by itself? This atheistic philosophy simply took its next logical step through Karl Marx, who imagined it applied to whole societies. Soon after, brutal dictators like

Joseph Stalin and Mao Zedong inflicted this philosophy on humanity.

Marxism held—and communism still holds today—to an idealized utopian society void of class struggles and therefore void of all evils. Like so many shallow sci-fi plots where the villain struggles to take over the world, some communists have seen utopia as a global communist society. As history has proven countless times, its proponents justify violent means to achieve the pinnacle of human evolution—a man-engineered society where no social class is lower than any other group. In their minds, man evolved from animals and thus bears no image of a God who does not exist. Those who reject God reject His Word. They have turned wrong to right in their struggle to imprison, starve, or slaughter anyone suspected of anti-communism. And why not, if indeed Genesis is just a myth and mankind is devalued to animal status? We get rid of rat infestations, so why not people infestations, too?

What did communist leaders have in common as they starved the countryside? They may have shared some personal elements in their backgrounds, but they certainly shared a low view of Genesis. When humanity constructs societies in the absence of core Genesis principles like creation, accountability to God, inherent sinfulness, families, stewardship of the creation, and universal morality, disaster soon follows.

On the other end of the spectrum, nations whose governmental structures tie back to those biblical principles seem to fare better. The United States, United Kingdom, and Israel, as examples, have enjoyed personal freedoms of individual citizens, reliable infrastructures, justice systems based on inherent human rights, growing economies, and pioneering scientific discoveries. One could argue that Genesis makes all the difference when it comes to building a society.

Genesis in Germany

> *"Darwin himself coolly predicted in* The Descent of Man *that the most highly developed humans would soon exterminate the other races because that is how natural selection works. Such casual references to genocide only began to seem reprehensible after Hitler, Stalin, and Mao demonstrated what they meant in practice."*[14]
> — *Phillip Johnson, 2000*

Nazism deserves its own section. Like the leaders of communism, Hitler rejected the authority of Genesis and used Darwinism to justify his Nazism. See chapter 6 for more on this comparison. For now, it should help to first consider the profound impact of Darwinism on entire cultures, including Nazi Germany not too long ago.

Much more than merely a statement on adaptation in biology, Darwinism has proven to be a total philosophy of life. Like man-made religions, it tries to answer where we came from, why we exist, and where we are going. Accordingly, we came from apes that came from fish, we exist to be the fittest or to propagate our genes, and we are going to oblivion. What better philosophy is there to undergird the full-on hedonism that we see in our culture today, where we try our best to vacuum all the fun we can get out of this short and supposedly pointless life?

Like communism, Hitler's Nazism was a logical extension of Darwinism into society. It ignores the fabric of real life, like individual rights that come from each individual's accountability and value to their Creator. Instead, individuals have only as much value as they can prove by being superior. To Nazis back then, and their philosophical descendants of today such as white supremacists, blond hair and blue eyes represented the master race of men. All others, being inferior (i.e., "less fit for survival"), have more in common with "lower" animals like apes. No wonder Nazis had no qualms about attempting to eradicate Jews and gypsies. No wonder they slaughtered Aboriginal Australians just to collect their skulls in attempts to manufacture evidence for human evolution.

In this way, "scientific" Darwinism flowed into social Darwinism. Nazis figured, as many today still do, that since nature has been selecting superior breeds or races of animals and humans for eons, it makes sense to join nature in this task. By killing the "unfit," Nazis believed they were participating in a grand and perfectly natural narrative of all life. As origins author Dr. Jerry Bergman wrote in a historical review of Darwinism's impact on Nazism, "Hitler believed that the human gene pool could be improved by using selective breeding similar to how farmers breed superior cattle strains."[15]

What did Nazis think about Genesis? Nazism, communism, and Darwinism share a similar view: God did not create this world. According to these worldviews, natural processes deserve all the credit. With a Creator God out of the picture, these atheistic philosophies fueled whole societies to slaughter tens of millions of people. Nazis rejected the Genesis assertion that man bears God's image. They felt no moral restraints when they starved and executed their victims. The Darwinian narrative justified the Nazi oppression of supposedly less fit human "animals" that bore no inherent dignity.

Many today doubt the Darwinian underpinnings of Nazi Germany despite the statements from Nazi leaders' journals and books. We would do well, therefore, to familiarize ourselves with what historian Richard Weikart brought together in his book *From Darwin to Hitler: Evolutionary Ethics, Eugenics, and Racism in Germany*.[16] Weikart does not argue that Darwinism inevitably causes Nazism but merely that as a matter of history, evolutionists at the time actively applied their evolutionary ethics to whole societies. German evolutionary propagandist Ernst Haeckel carried Darwin's ideas into elite German circles, and they eventually wound their way to Hitler and then across the whole country.[17] Weikart clearly traced anti-Genesis concepts like the inequality of peoples, moral relativism, and animal ancestry from Darwin to Hitler. If people groups—as merely evolved apes—cause problems, why not just wipe them out like one would a plague of frogs?

This could possibly oversimplify matters, but for the sake of clarity we might say evolutionary ethics replaced the Ten Commandments with new ones: thou shalt obey the laws of nature, including the laws of selection. Nazis believed they not just could but *should* de-select the unfit. By killing off the physically or mentally crippled, Nazis felt they were performing their moral duty to improve mankind's fitness. Same goes for old people. Scientists used the term eugenics perhaps to cover the horror of mass murdering the aged.[18] Darwin wrote in his biography:

> A man who has no assured and ever present belief in the existence of a personal God or of a future existence with retribution and reward, can have for his rule of life, as far as I can see, only to follow those impulses and instincts which are the strongest or which seem to him the best ones.[19]

Nazis borrowed this relativism to construct their own pro-Aryan, anti-Genesis set of rules.

Any problems with this, other than the fact that evolutionary ethics ignore the Bible, including Genesis? Sure. For one, no scientist has yet published results of a study that has shown any natural process that improves a creature by adding new and useful information. Instead, creatures either overcome environmental challenges, often through internally structured adaptations, or they don't. So, what if certain bacteria adjust their metabolism to grow faster in a test tube? They remain the same kind.[20] Second, Nazi's definition of "inferior" was entirely subjective. Who is to say that blonds are superior except for those with blond hair? Indeed, some black Americans today flip the same script, claiming superiority through their ancient Egyptian ancestors.

Like communist China and Russia, Nazi Germany should have shown all survivors of those regimes and their descendants that attempts to live life without Genesis and societies that reject God and His Word fall off a cliff. This suggests a lesson from history for us: respect Genesis and live by its basic principles, like God being the Author of life and morals and man, and society can soar and not sink.

Genesis and Hot Buttons

Sodomy

The Bible confronts same-gender sexual behaviors alongside other unholy sexual acts, like adultery or premarital sex. Scripture calls it sodomy when we violate the intended function for sexual body parts in this way. Bible translators chose the word "sodomy" to translate the Greek word *arsenokoites*, which refers to a male lying with another male as with a female. The name hearkens to that ancient city the Lord judged with fire according to the records in Genesis 18 and 19.

Logically, homosexual and other unholy sexual behaviors flourish with a rejection of Genesis origins. Once we convert to legend the judgment on Sodom and Gomorrah, the creation of Adam, and the creation of the first married couple as a union of man and woman as opposed to man and anything else, then anything goes. As part of this popular trend of converting history to legend and then legend to faraway fancy, some cite a lack of archaeological evidence for Sodom and Gomorrah. However, evidence exists from those remains, and archaeology of candidate sites deserves a peek. What's been found from the ground corroborates the testimony of Jesus, who referenced the Sodom events as though they really happened (Matthew 10:15).

As a preface, we know that even conservative archaeologists still debate about which precise site was Sodom. Perhaps someday they will set apart one site as a standout, even as has happened with placing the biblical Shiloh at today's Khirbet el-Maqatir, the biblical city of David within today's Jerusalem, and the biblical Mount Ebal at its traditionally named location. These and dozens of other sites have required decades of deciphering. Thus, the following description of a candidate site for Sodom may not end up being the exact site, but for now it at minimum illustrates that quality candidates exist and at most illustrates that quality candidates are worth serious thought.

The five (Genesis 14:2) cities of the plain (Genesis 19:29) including Sodom would have lain near the once-lush Jordan River valley's broad plain. Archaeologists with a low view of Scripture seem quick to discard candidate sites for the five cities. However, some with a high view of Scripture—who treat Genesis as history—have made compelling identifications.

Along the southwest side of the Dead Sea, five now-dried creeks would have supplied water that flowed in wetter times down to the plains from eastern uplands in modern-day Jordan. Although problems remain with its timing (and what else is new in the world of archaeology?), the geography of Sodom matches the location of modern Bab edh-Dhra, and Gomorrah sizes up as today's Numeira, the biblical Zoar (with its even older name Bela, from Genesis 14:8). Archaeological sites at Safi, Feifa, and Khanazir lie even farther south to round out the five cities of the plain.[21] Each had its own stream that once flowed east through now-dry valleys that incise the Jordanian Highlands.

Genesis 14:10 says, "Now the Valley of Siddim was full of asphalt pits." Half of Bab edh-Dhra now perches about 90 feet higher than the other half of the ancient city. A massive earthquake split the city along the Jordan Rift Valley fault. The event would have ignited underground asphalt to incinerate the whole valley. Thus, Abraham "looked toward Sodom and Gomorrah, and toward all the land of the plain; and he saw, and behold, the smoke of the land which went up like the smoke of a furnace" (Genesis 19:28). And yes, Abraham's vantage from Mamre over 30 miles away fits the geography. Even today asphalt oozes from nearby rocks. Those willing to see it can recognize geology and archaeology that support Sodom.

The Lord Jesus in Luke 17 compared the suddenness of Sodom's destruction to the suddenness of His coming judgment. He, plus Paul, Peter, Jude, and John, believed Sodom was a real place that had earned its own undoing. To pull this miraculous destruction from the realm of myth into human history implies that God really cared enough about protecting the surrounding peoples from Sodom's evil influence that He destroyed the five villages. He also cared enough to protect His own when He "delivered righteous Lot" (2 Peter 2:7) from that judgment. The Bible paints a picture of the Creator's disapproval of sodomy and the Sodomites' other sins noted in Ezekiel 16:49-50. He wouldn't have warned us if He didn't care about us! His care comes through in His instruction that the gospel is for anyone caught in any sin. Indeed, God provided His laws not for us to add to our to-do lists but for us to see how desperately we should desire His forgiveness. Paul wrote:

> But we know that the law is good if one uses it lawfully, knowing this: that the law is not made for a righteous person, but for the lawless and insubordinate, for the ungodly and for sinners, for the unholy and profane, for murderers of fathers and murderers of mothers, for manslayers, for fornicators, for sodomites, for kidnappers, for liars, for perjurers, and if there is any other thing that is contrary to sound doctrine, according to the glorious gospel of the blessed God which was committed to my trust. (1 Timothy 1:8-11)

What is this gospel, what does Genesis have to do with it, and does belief in Genesis as history relate to the sodomy of our day? The gospel is the good news that no matter what sins we have wrapped ourselves in, the Lord can rescue us right now from their penalty and will eventually save us from their very presence. It is the simple and true message that in the moment when we turn from sin and cry out to God, He grants forgiveness and everlasting life. This message includes the concepts of sin as real wrongdoing, God as a real judge who will enforce a universal moral law, and God as the One to whom we must give account as our Creator. In other words, this gospel message that the Lord offers to all us sinners assumes truths about God's character that tie back to Genesis. This glorious gospel also asserts truths about God's mercy on the guilty and grace for the humble.

Societies that jettison Genesis reject God as its divine Author as well as God as special Creator. Without this Creator, we pretend we have no divine Judge and deceive ourselves into thinking our sins are not wrong after all—or at least not very bad. And there goes any perceived need for a redeemer or savior. When unchecked, this rejection sears countless individual consciences and points whole societies toward degrees of wickedness that have compelled the Lord to remove them from Earth. The steps He took back then helped give then-future generations like yours and mine a chance to exist—to see and know Him. He removed Sodom and Gomorrah, Israel temporarily (Amos 4:11), Moab (Zephaniah 2:9), and very likely other nations not even named in Scripture. Genesis contains the history behind the doctrine that God did not make husbands to marry men or wives to wed women, but husbands for their precious brides. Rejection of the basic principles found in Genesis, such as marriage between man and woman, is a big deal.

The Lord later explained His dealings with Sodom as He blew the whistle on wayward Judah, saying:

> "Look, this was the iniquity of your sister Sodom: She and her daughter had pride, fullness of food, and abundance of idleness; neither did she strengthen the hand of the poor and needy. And they were haughty and committed abomination before Me; therefore I took them away as I saw fit." (Ezekiel 16:49-50)

What has this to do with the book of beginnings? Just that to be proud and haughty means to raise one's fist against God's authority over us—the authority that He rightly has by virtue of making us out of dust. This rebellion can take the form of ignoring the poor and needy, committing sodomy, or worshiping ourselves in countless other ways.

Some speak of how our genetics determine or at least sway our sexual orientation. Would God punish Sodom and the five cities for genetics that He gave them? He would never! According to Genesis and the other Scriptures, they—not their genes—chose proud attitudes and wicked deeds. In accord with this, an August 2019 study of 477,522 individuals disproved this "gay gene" concept.[22] One of its coauthors confirmed that with this many people included, there no longer remains any doubt that there's no genetic signal that tells if an individual will turn out gay. We already knew that based on identical twins who make either choice, but now genetics confirms Genesis in yet another realm.

In this section we have considered what amounts to a negative message: it's shameful for sodomites or any other sinner, including me, to reject the God of Genesis. But the goodness of God's grace and love toward sinners can overwhelm any who seek Him. I have rejoiced with homosexuals who found forever forgiveness through the Lord Jesus Christ. I will see them in that promised garden, and I aim to see more who turn to their Savior.

Gender Selection

A news headline on March 31, 2021, read, "CNN Claims There's No Way To Tell If A Baby Is A Boy Or A Girl."[23] Indeed, the CNN news article to which it referred said, "It's not possible to know a person's gender identity at birth." I don't know how many maternity wards the author, Devan Cole, has visited, but just one or two stops there would showcase yet again humans' binary gender setup. Cole represents a growing number of others in our increasingly twisted society who argue that biological sex is "a disputed term."[24] Of course it is, but all truths are disputed by those bent on denying them. Despite protestations to the contrary, mothers keep giving birth to baby boys or girls.

Some public schools now teach young children they can choose if they want to be a boy or a girl, and some parents back up the message at home. Transgenderism has gone wild. Why? Perhaps widespread openness to sexual perversions combined with relativism helped pave the way.[25] Certainly the idea that we are all just animals anyway paves a wide path toward gender confusion, and this stands firmly opposed to our miraculous creation as male and female in the Garden of Eden as revealed in Genesis. But can we really choose our gender? Surgeries and hormone doses can help make a female look more like a male or vice versa, but nobody can totally avoid the gender they were born with.

In addition to redefining gender and assuming gender is a choice, a woman named Kenny Jones offered a third example of gender confusion in our culture. She transformed to a male, then complained of feeling dysphoria over having to buy feminine products that looked so "feminine." Proctor & Gamble then removed feminine images from their feminine product line so as not to offend protestors like Kenny who feel such products should acknowledge that it's not just females who "can experience menstruation."[26] Medical doctor and ICR President Randy Guliuzza noted that men menstruating "has never been observed." He called these concepts "flights of fancy" and "utterly foolish."[27]

Claims that we cannot know a person's gender at birth, that teachers should tell their tiny students to choose their gender, or that men can menstruate all ignore anatomy, and they share a common cause. Now more than ever, we have to get back to basics. And that's what motivates me to write this book. Where did genders come from? What may lie behind these foolish and destructive gender claims? That good, old Bible has good answers.

So many people forget or ignore that humans have taken male and female forms since creation. Genesis 1:27 tells us why and when our gender began. It says, "So God created man in His own image; in the image of God He created him; male and female He created them." Why male and female and not just male or female? We may not know the answer until the next age, but we are sure that the Lord wrote it this way for a specific reason. The male and female referenced were the world's first married couple. In this context, perhaps that couple bore "the image of God" in that a husband and wife—who are different persons with different essences—can share intimate fellowship like that of the Father, Son, and Spirit.[28] In this way, married couples might be living metaphors of the three Persons who, unlike a husband and wife, share the same essence. Our culture has largely forgotten all this. We no longer know where we came from, so it's no wonder, really, that we don't know the difference between male and female or right and wrong. We call Genesis "junk," claim the Bible is bunk, and shove the Creator further away. What has infected us that we feel so comfortable denying undeniable facts like genders? The answer must lie in our rebellion.

Isaiah's godly assessment of his culture resonates today:

> For you have trusted in your wickedness
> You have said, "No one sees me"
> Your wisdom and your knowledge have warped you
> And you have said in your heart, "I am, and there is no one else besides me." (Isaiah 47:10)

But like Isaiah's countrymen did, ours can see wisdom restored! How? How about we beg the Lord to straighten our warps—first those in our own hearts, and then those in our neighbors'. As Isaiah directed them, so we too need desperately to return "to the law and to the testimony!" (Isaiah 8:20). Return to the Word of the Lord! To His record of our beginnings! To the God of the garden of which that record speaks! To the One we have offended! The degree to which we reject our own rebellion and humbly request the Lord Jesus' help will establish the degree of harmony between our well-designed genders even in this life.

Abortion

Abortion is personal and painful, so I want to be sensitive and brief, but we must discuss it since it has logical ties to Genesis. The term abortion refers to the removal of a developing baby from his or her mother's womb. While some call this murder, others don't give it much thought. What ideas underlie the stance that a young life hidden inside a belly has no value but that the same young life hidden inside a bassinet some months or minutes later is suddenly worth protecting?

Rejection of the God of the garden involves the rejection of humankind bearing the image of God. Indeed, Western societies routinely teach that humans emerged from ape-like ancestors and not at all from a Creator. If we are all just hairless apes, then the removal of those deemed a burden, including the most defenseless demographic of all, is reduced to the level of eradicating mosquitoes.

In contrast, if a new human life bears the image of God, then each baby carries a high value indeed. Other Bible passages back this view, like 2 Samuel, which tells us that King David said about his dead newborn, "I shall go to him" (2 Samuel 12:23). Our great Creator so loves even the unborn that He will reunite them with their parents.

Rather than having to give an account to God for taking His children, God-fearers can learn from Genesis why they should cherish and defend them. The Lord Jesus shared this high view of persons when He said, "Do not fear therefore; you are of more value than many sparrows" (Matthew 10:31). New life has inherent value because God created and loves life. That baby is an expression of His very nature and image.

Over 18 of 100 pregnancies in the United States ended in abortion in 2017,[29] and other years carry similar numbers such that tens of millions of Americans have aborted their babies since the famous 1973 *Roe v. Wade* court decision.[30] How much of this carnage can we honestly expect the Lord to endure? He does have a limit. We do need to return to Him in humility and faith that He will heal and forgive.

I'm sure someone could link abortion trends to church attendance, and church attendance to respect for Genesis. But the logic that ties respect for life to respect for the Maker of life (and even His record of a tree of life, Genesis 2:9) in the beginning needs no such analysis. Creation according to Genesis replaces man-made nonsense about ape origins with the truth that God handcrafted the first humans to be fruitful and multiply generations of bearers of God's image. It lays the logical and historical facts that establish the value of our little images of the glorious Creator.

Collateral Damage

How does our view of humanity impact the way our societies wage war? Genesis informs our view of, and value for, humanity. It teaches that humans are made in the image of God. This means that humankind reflects God in a variety of ways such as our creativity, reasoning abilities, capacity for experiencing and expressing emotions, and especially our spiritual capability to know God and love others.

Genesis 1:26 says, "Then God said, 'Let Us make man in Our image, according to Our likeness.'" The English word "image" translates the Hebrew word *tselem*. Elsewhere in Scripture this word is used to refer to idols molded to represent people's imaginary gods. If one carries a 3-D model concept from this most common use of *tselem* backwards to Genesis 1, then along with the nonphysical similarities between God and humans—all of which God designed in order for us to know Him—our physical bodies may also resemble the physical body Jesus knew He would take upon His incarnation. He also took this form in His pre-incarnate appearances, for "the LORD God [walked] in the garden in the cool of the day."[31] One cannot walk without legs.

So, when it comes to war, we see nations with a high view of Genesis also carrying a high view of humanity. Cultures with Genesis in their history, generally speaking, have laws that prescribe the protection of noncombatants during wartime. Of course, the opposite generally holds, too, when cultures with a low view of Genesis—in particular Genesis 1, which conveys the idea that men and women contain inherent value as special creations who both bear God's own image—slaughter innocents with disregard during battle. Most of us have heard the phrase that ideas have consequences, and it surely holds true. Is it any wonder, therefore, that a faction of Islamists who hold most tightly to the teachings of the Qur'an and of Muslim prophets rather than of Genesis did not simply ignore noncombatants during 9/11 and so many other attacks but even targeted them?

The late Mohamad Jawad Chirri said, "Muslims do not subscribe to the contents of the first chapter in Genesis book because it shows some discrepancies."[32] Those supposed discrepancies reduce to mere apparent discrepancies, resolve upon closer inspection, and leave the first chapter of Genesis perfectly intact, as I argue elsewhere in this book. The point here is to acknowledge a logical connection between violent Islamists and their rejection of Genesis.

Other nations, including many from Europe, have agreed to Protocol I, a 1977 amendment to the Geneva Convention that outlaws indiscriminate attacks on civilian populations. Article 51.2 says, "The civilian population as such, as well as individual civilians, shall not be the object of attack."[33] These nations generally carry histories of Christian respect for the Bible's first book.

Animal Rights

No attempt to address how Genesis impacts culture would seem complete without mentioning those folks intent on making animals into persons. Lawyers from the Nonhuman Rights Project, for example, have appointed themselves representatives of Happy, an elephant at the Bronx Zoo. According to *The Guardian*, this "animal rights group hopes it will effect a legal breakthrough that will elevate the status of elephants, which…should have the fundamental right to liberty."[34] Its leader Steven Wise argued in his Supreme Court case that since Happy is a person, the Bronx Zoo detains her illegally. The group had been unsuccessful in using the judicial system to grant personhood to Kid and Tommy, two captive chimps. What societal beliefs have contributed to this—the latest in a string of attempts to totally redefine what it means to be a *person*?

The Nonhuman Rights Project website says of its organization, "We use the term *nonhuman rights* to remind people that human beings are also animals—the only animals with legally recognized and enforceable rights."[35] Well, there it is. They say humans are animals, not made in the image of God as Genesis says, but instead made by natural forces.

But here's an irony. The very Book that these people disdain supplies the rational basis for treating animals (and humans, for that matter) with respect. If Happy, Kid, Tommy, and Steven Wise all emerged spontaneously from stardust instead of from a Creator, then on what basis can this group rest their moral case? Stardust has no morals. Neither does Happy have any sense of violating God's laws or needing salvation from rebellion against its Creator. People and only people have that sense because that is part of what it means to be made in God's image. God does have morals, and He introduced the first humans to the reality of right and wrong right there in the beginning.

Climate Change

Just last week I actually read one of those annoying pop-up ads that obstructs web pages. It began by saying that climate change is the defining issue of our time. It then described the company's commitment to keeping our environment clean. Who doesn't favor keeping Earth clean? On the other hand, what evidence backs up the claim that climate change is a more defining issue than abortion, terrorism, apathy, or godlessness? Climate change hysteria has reached the point that 11,000 scientists have signed a declaration of emergency that calls for worldwide population control.[36] What evidence demonstrates that humans are the root of this reportedly emergency state? Does reliable, repeatable science underpin such a dire decree?

Certain climate science observations have eroded my confidence in climate alarmism—the idea that climate is humanity's number one problem. First, global temperature rose during most of the 20th century. So did CO_2. So far, we agree. But the CO_2 began rising *after* the temperatures. Effects are supposed to follow after their causes. How could CO_2 cause a temperature rise if it did not rise first? It instead appears that increased heat released more CO_2 into the atmosphere. If this is the case, then we should not point to CO_2 as the cause of warming trends.

The second climate-related observation that adds to my skepticism toward climate alarmism is that global temperatures did rise, but they leveled off during the 1990s and 2000s.[37] All the while, industrial emissions were increasing. I was told that emissions were supposed to be heating the globe. How can increased emissions be the sole cause of global warming if they failed to cause warming for a decade?

Models of climate change that have solar activity as the main driver of climate change make the most sense of the available data. But a line like that offers no leverage to rationalize political power grabs. A primarily sun-responsive Earth atmosphere may sound boring, and it may oppose climate extremism, but plenty of silenced scientists agree with it strongly enough for them to risk their reputations and jobs.

Those who wish governments would force supposed climate-saving behaviors on all peoples ought to first explain the Medieval Warm Period (MWP). It occurred during the Viking Age, when Earth's global temperature compared closely with today's warm period. Although climate alarmist researchers massage the temperature proxy data to insist that the MWP never happened, England and Germany grew grapes back then, lost that industry in the Little Ice Age that peaked around 1800, but now can make wine again.

Did the Vikings pump factory emissions into the ancient atmosphere that long before the Industrial Revolution? Of course not. Something other than humanity seems to drive climate trends. There's more, but these three observations—out-of-place CO_2, steady temperature in the 1990s, and the MWP—seem sufficient to reveal how nonscientific this supposed defining issue of our time really is. If science doesn't lie at the heart of climate alarmists, then something else does.

Could our nature-only dogma play a role? Public school textbooks teach that debris collisions in space naturally formed a molten earth billions of years ago. This comes despite the overwhelming observations of space rock collisions that smash apart instead of build up. In contrast to this hot, molten, imaginary natural beginning, the Bible indicates watery, supernatural origins. Genesis 1:1 says, "In the beginning God created the heavens and the earth." Not collisions—God.

Evolutionary theorists imagine a molten earth that must have taken eons to cool down enough to hold its water. But then they have the problem of getting all the water we have onto Earth from outer space. They have no defensible idea of how Earth received its water. Models of impactors that carry water have difficulty answering the question of how heat from the impact doesn't vaporize the water it just brought in.

Thus, both science and Scripture offer reasons to doubt the assumption of a molten original earth. But that's what textbooks have taught generations of students. This anti-miracle bias puts forward a molten earth that supposedly evolved an atmosphere so frail that we humans have the potential to send it back to the oblivion from whence it supposedly came.

A high view of Genesis, and two of its verses in particular, erases such fears. The first verse noted above teaches that God made Earth. He must therefore have a purpose and plan for all that happens upon its surface. It also implies that the climate is generally stable. No wonder we find biological systems like ocean algae—systems that climate models largely ignore—that maintain a constant balance of atmospheric gases including CO_2.[38]

The second key climate verse accompanies a solemn promise that the Lord made with the earth's animal and human inhabitants for all time. It says, "While the earth remains, seedtime and harvest, cold and heat, winter and summer, and day and night shall not cease" (Genesis 8:22). This promises that God will maintain the earth's suitability for farming forever.

Does God's promised regularity mean that Christians should promote the idea that humankind can wreck the very climate that our God promised He would sustain and keep stable? Not at all. Christians should care about the earth because God has given humans the responsibility to steward His creation well. The last verse in the book of Jonah helps reveal the Lord's care for His whole creation, saying, "And should I not pity Nineveh, that great city, in which are more than one hundred and twenty thousand persons who cannot discern between their right hand and their left—and much livestock?" (Jonah 4:11).

My point is this: ignorance of Genesis—whether accidental or willful—plays a role in cultures that advocate such drastic measures as curtailing whole economies or populations in the name of saving the planet. Such measures reduce to our sinful insistence on saving ourselves instead of humbly pursuing the real Savior. A return to Genesis would transform the way we think. It would replace the perceived fragility of an earth born from accidents with the surety of an earth spoken into existence and maintained by that same word until God replaces it.

> For this they willfully forget: that by the word of God the heavens were of old, and the earth standing out of water and in the water, by which the world that then existed perished, being flooded with water. But the heavens and the earth which are now preserved by the same word, are reserved for fire until the day of judgment and perdition of ungodly men....Nevertheless we, according to His promise, look for new heavens and a new earth in which righteousness dwells. (2 Peter 3:5-7, 13)

Living in light of Genesis and its loving Lord would restore His rightful place in our minds as Ruler of the universe, of Earth's climate, of our perceptions of our bodies, and ultimately of each of our deepest disappointments and dreams.

Genesis and America

In a chapter titled "Culture Without Genesis" it seems reasonable to contrast the origins of those cultures that rejected God's Book with its record of beginnings to cultures founded by people who respected His record. Many cultures and countries have waxed upon the surface of Earth as their respect for Genesis stays high, only for them to blink out generations later after its peoples oppose God and His Word. One culture that seems to have begun with a high regard for Genesis and now abandons it is the United States of America. I don't want to leave the impression that America or any other country is perfect or has no problems. Far from it! But I do find a logical connection between the way people groups respect the God of Genesis and their success and happiness.

The U.S. Pledge of Allegiance includes the words "one nation under God, indivisible, with liberty and justice for all." Under God? How did God make His way into the national pledge? For one, He introduced Himself to the Puritans during the Great Awakening. Back then, the Lord captured the hopes, imaginations, and the very hearts of influential American men and women. The next generation emerged with a high view of God, a high view of His Word, and, in short, a worldview that treated Genesis as history. Their descendants became the founding fathers of the U.S.A.

I happen to be in Boston as I write this, passing through on a speaking trip. Even though my schedule keeps me from being able to visit many historical sites, I can't help but absorb the sense that Boston retains the spirit of America. When modern Boston began, the clergy dictated church terms and procedures. In contrast, many founding Bostonians were Congregationalists who had active voices on processes and issues that affected their own church bodies. This arose from and propagated a biblical worldview that shaped America.

The concept of each person holding a valid voice carried forward into the American culture and government. Instead of all obeying a king's edict, today's ballot boxes trace their history to Congregationalism. And what gave our American forebears the idea that the people should have a say in their governance? Many of us today agree that, in the wonderful words of *Horton Hears a Who*, "A person's a person, no matter how small."[39] After all, each person carries God's own image according to the Genesis that Americans once held high.

1920 saw the 19th Amendment to the U.S. Constitution give women the right to vote. Where did Americans find the idea that women should play a direct role in politics? It took a while for suffrage efforts to bear this fruit, but how can we deny that its key foundation—the equality in value if not in stature—appears in Genesis 1:27? Both men and women bear the image of God. The whole of Scripture agrees, of course. For example, Peter wrote that husbands should treat their wives with understanding and honor (1 Peter 3:7).

A high view of Genesis surely inspired the words in the Declaration of Independence "all men are created equal." Hindu beliefs reject that. Its ancient caste system treats certain folks as haves and others as have-nots. This non-Genesis inequality has produced generations of lower caste members who have suffered all the worst ills of extreme poverty and abject disdain. Who can reasonably deny the Genesis foundation for the core concept that all peoples should be treated with respect simply because all equally bear the image of God?

William Wilberforce, a key influencer in the United Kingdom, became passionate and determined to fight slavery. What drove him? Some of the final lines that he used in his abolition address to the House of Commons on Tuesday, May 12, 1789, hearkened to a Genesis foundation. He said,

> I could not therefore help, distrusting the arguments of those who insisted that the plundering of Africa was necessary for the cultivation of the West-Indies. I could not believe that the same Being who forbids rapine [violent land seizure] and bloodshed, had made rapine and bloodshed necessary to the well-being of any part of His universe.[40]

I do not wish to offer up too lofty a view of my country simply because it is the one into which I was born. Many countries have made critical changes that led to various righteous equalities. I simply wish to affirm how a high view of Genesis elevates a society's view of humanity. Paul the apostle confirmed the Genesis perspective when he told the crowd in Athens, "And He has made from one blood every nation of men to dwell on all the face of the earth, and has determined their preappointed times and the boundaries of their dwellings, so that they should seek the Lord" (Acts 17:26-27). Many places around the globe can trace these societal perspectives back to Christian roots. And Christianity's branches run down to its Genesis roots.

Culture Minus, or Plus, Genesis

Let me attempt to summarize the reason for this chapter. I wanted to explore the idea that, over the long haul and generally speaking, cultures that jettison Genesis (alongside the God whose Word it is) begin trying God's patience. He is extraordinarily patient. Thank God for His patience with those who hate Him. If He were not so, then you and I would not have the opportunity to be born or to know and enjoy Him forever as we do.

But God does blow the whistle at a certain point. He did it on the whole world before the Flood. He did it at Sodom and Gomorrah. He blew the whistle on Egypt, Canaan, Moab, Assyria, Babylon, Persia, and even ancient Israel, which He is in the process of resurrecting out of the nations. So, that's the negative but necessary message of a chapter titled "Culture Without Genesis." Individuals comprise cultures. How many individuals prove Proverbs 29:1, "He who is often rebuked, and hardens his neck, will suddenly be destroyed, and that without remedy"? If enough people within a culture stiffen their necks, then there goes the neighborhood.

Our present observation, seen from another angle, is that one need merely study history to help decide how to treat Genesis. See how well it went for those who rejected, as compared to those who embraced, the Bible's history that included the lost garden that God promises to remake even better. Culture minus Genesis ends badly. But a culture that embraces God's Word finds a future. Those who live in it have a chance to reorient their hearts toward the truths about where we came from and why we are here—in short, a chance to live in light of Genesis.

5
GENESIS AND THE BIBLE

"And beginning at Moses and all the Prophets, He expounded to them in all the Scriptures the things concerning Himself."[1]
(Luke 24:25)

Five Core Concepts That Lean on Genesis

In my journey from evolution to creation, I inevitably faced the Bible itself. Where do its New Testament doctrines derive their force and function from? How did the prophets treat Genesis? How much of what we practice in daily life—getting dressed, getting married, getting out of jail—can we justify with Scripture? In short, if those who followed the Lord Jesus before me—who longed to be reunited with their long-separated Savior according to His promises—treated Genesis as history, then who was I to treat Genesis as anything less? I began to take notes on how closely some basic Bible concepts are tied to the book of beginnings.

Marriage

According to Genesis, God designed a man and woman to become one in marriage. We become one physically in the act of marriage, and that can bring forth children who have their own opportunities to know and enjoy the Lord during the next generation. We become one spiritually by sharing one another's joys and struggles and by submitting to the Lord together, regularly. We become one intellectually as we question and listen to one another's beliefs, thoughts, and attitudes. All this begins with a relationship based not on blood but a commitment—the marriage covenant.

Surely this pictures believers' relationship with the living God. We become one with God based on His covenant with us. He covenants with those who repent of sin and trust Christ to save them from the penalty of their sins now, the power of sin over believers' lives as they walk with Him (Colossians 2:6), and someday from the very presence of sin. We simply say, "I will follow You," and all these benefits accrue. Similarly, in marriage the husband covenants to love his wife even when she sins against him, and she covenants to respect her husband even when he sins against her. And each party simply says, "I will." That is to say, it's a simple commitment to express even though it is very difficult to carry out.

Marriage pictures how humans rightly relate with God. Marriage also forms the foundation of human society. But marriages are crumbling, as are families—many of which undoubtedly have a low view of Genesis. Studies that investigate this issue consistently confirm that kids without two traditional parents have a higher likelihood to fail in school, commit crimes, find their way to drugs, earn lower average incomes, and perpetuate their miseries to the next generation.

Those same studies are catching up to Genesis as they reveal that children raised in traditional, two-parent homes—even non-Christian homes—experience more academic achievements and fewer crimes, drugs, and infidelities. The government simply cannot achieve what God designed a family made of a husband and a wife to accomplish. It's like trying to use a bowling ball instead of a scalpel to perform brain surgery.

Is it any wonder that today's cultures, like urbanites from Denmark whose marriage rates have plummeted, also have a low view of Genesis? Or that Genesis plays a low or no role in the lives of so many who perpetuate broken families by flitting from bed to bed without regard for the children they create? Our world desperately needs to rediscover the virtues, blessings, joys, and benefits of biblical marriage and of its pure foundation in the garden.

It took me a while to appreciate the origin of marriage. I still remember the preacher's Southern voice and lilting intonation as he said, "God invented marriage. It was His idea." Sounds simple to those who grew up hearing it, but it came as news to me. Once I learned about the first marriage in Genesis, radical changes began.

I suddenly started to think of marriage as a worthwhile institution instead of a setup for failure and thus something to be avoided. Not many months later, God blessed me with a magnificent wife, right out of the blue. Soon enough, five babies emerged in four short years. How, you ask? Twins. My youngest child has now graduated from high school with a smorgasbord of achievements and a passion for the gospel. God gets the credit—He started it and enabled every positive step.

Whatever positive influence my bride and I had on our kids started with work on our relationship with one another—and that began with Genesis. The book of beginnings, for example, taught us that we are to become one. For us that meant giving up individual preferences for the other and becoming deliberate in meeting each other's needs. We actually enjoyed our kids' teenage years as we tried to disciple them in the Lord and walk with them through life's challenges and joys. Today they are, for the most part, embracing faith in Christ, all because of God's grace. And it all started when the Lord taught us the origin of marriage in Genesis.

Our experience at a single-family level has convinced us that living in light of Genesis would elevate a whole society's value of marriage. We would love to see a positive, culture-wide impact of marriage built on the ideals of its Inventor.

Clothing

Isn't it nice that we wear clothing? To anyone whose conscience is not seared, a world full of naked humanity sounds creepy to the max. Christians are supposed to be transparent with one another, but not that transparent. Some things need to remain hidden. Of course, animals don't clothe themselves, only men and women. And we do so because of what happened in Genesis 3:21: "Also for Adam and his wife the LORD God made tunics of skin, and clothed them." In His grace, God covered their shame. He did that, at least in part, to be a perpetual reminder that our sin requires a covering. God provided His only Son as the ultimate covering—a covering so complete that it stands in the way of God's righteous judgment for those who repent of sins and trust

Christ. Clothing comes from Genesis. If Genesis is a bunch of myths and if we arose from apes, then why do we wear clothes?

Creating

A third core concept that reaches back to Genesis involves uniquely human creativity. Have you ever felt that moment of satisfaction after completing a project? Maybe you finished that essay, performed a song, followed a recipe to perfection, finished a race, or built a porch. Ecclesiastes 5:19 expresses this human experience when it says, "As for every man to whom God has given riches and wealth, and given him power to eat of it, to receive his heritage and rejoice in his labor—this is the gift of God." When we rejoice in the fruits of our labor, do we not experience a shadow of what God experiences when He makes something out of nothing? And I suspect this ability—and our ability to reflect on this ability—folds into what it means to be made in the image of God inasmuch as we can please our heavenly Father with our efforts like today's children please their earthly parents with flawed but unique creations. One difference between our and God's creativity is that although we can craft paintings and motorcycles with raw materials, God can craft anything with no materials. Still, we both craft things.

Why do we get married, wear clothes, and build cool stuff? Because Genesis teaches us to become one in marriage, Genesis teaches us that our sin makes us ashamed to be naked, and Genesis teaches us that we are made in the image of a creative God. The Lord even watched to see what names Adam would invent for those animals God brought to him.

The apostle Paul extends this creativity to the gospel. I had the privilege once of meeting Pastor John MacArthur. Like every good fan should, I asked him to sign my copy of his study Bible. Alongside his signature he wrote, "2 Cor. 4:6." This verse says, "For it is the God who commanded light to shine out of darkness, who has shone in our hearts to give the light of the knowledge of the glory of God in the face of Jesus Christ." Here Paul draws a Genesis parallel. Any God with enough power to command light to exist when the moment beforehand there was no such thing as light also has enough power to command a new you to come out of nowhere and from nothing but Him. God is the Creator of all things—this much Genesis makes clear. And 2 Corinthians makes it equally clear that this same Creator God can create all things new within the heart of anyone who calls on Him. Have you called on Him in this way?

Scripture

Now for a fourth core teaching that springs from Genesis. One of the first verses that jumped off the page during my discipleship was 2 Timothy 3:16. I soon memorized it. It says, "All Scripture is given by inspiration of God, and is profitable for doctrine, for reproof, for correction, for instruction in righteousness." Without a reliable way to access God's thoughts, how could we know them? We would remain forever in the dark.

God solved that problem. The Bible itself says it is God's Word. It authenticates that idea with superhuman internal features like perfect internal consistency, a Savior-focused theme that spans all 66 books, and precisely fulfilled prophecies. It even describes real human nature with no hidden agendas. It exposes not just the virtues but the deep flaws of those men and women who walked with God through the ages. If the Bible alone contains the truth about how sin started, why offending God is serious, and God's plan to rescue sinners, then humankind desperately depends on it as the singular source of God's words.

What does this have to do with Genesis? Well, the core teaching here is that the Bible is the Word of God. And Genesis definitely belongs in the Bible. The prophets and apostles who penned the Old and New Testaments and the Lord Jesus, toward whom both pointed, referred to the words in Genesis as Scripture. Remember 2 Timothy 3:16? It does not say "most Scripture" or "some Scripture is given by inspiration of God." It says "all Scripture." As soon as we claim that a certain verse was not given by inspiration of God or that a certain verse's words do not mean what they usually mean in their normal contexts, we dim our own access to the light of God's very Word. Each verse that we erase blocks another ray of light that God intended for us to know, cherish, defend, learn from, and live by. On the other hand, each verse we study illuminates more of God's character. When we dig deeper, study, and apply Scripture's words, verses, and passages, then God deepens our relationship with Him through His Son the Messiah.

Where did the whole concept of the Bible being the Word of God come from? Genesis. The phrase "God said" occurs 10 times in chapter 1 alone. Genesis 5:1 says, "This is the book of the genealogy of Adam." The word "book" here refers to words written down. Words can come pressed into clay tablets, inked onto papyrus, carved into rocks, or written with ink on animal skin parchment. Today we may use ink on paper or keystrokes on smart phones. The means don't matter. The words do. And this verse plus other Bible clues suggest that words—indeed, the first words from God—were there in the beginning. After all, "in the beginning was the Word, and the Word was with God, and the Word was God. He was in the beginning with God. All things were made through Him....And the Word became flesh and dwelt among us" (John 1:1-3, 14). The Bible's doctrine of the Bible starts with Genesis.

Death Before Sin?

Sin produces death and this is the main problem that Jesus came to solve. How do we know? This news reaches us from the ancient text of Genesis. Why do we die? "And the LORD God commanded the man, saying, 'Of every tree of the garden you may freely eat; but of the tree of the knowledge of good and evil you shall not eat, for in the day that you eat of it you shall surely die'" (Genesis 2:16-17). And the first man and woman ate. Death then entered the universe, and they eventually, but certainly, died.

In stark contrast, evolutionary history holds that death has existed for billions of years. More than that, it treats death of the unfit as the means by which new creatures arise. As such, evolutionary history, mistakenly taught as factual science, flatly contradicts the clear history given in Genesis. If all things, including death, continue as they were from the beginning instead of after sin, then what sense does it make for Jesus to have taken upon Himself our just death penalty? If death is not the consequence of sin but was around long before it, then death would not be the enemy that Jesus defeated at His resurrection.

These five core teachings all stem from Genesis: marriage, clothing, creation, the Bible, and sin. If we abandon Genesis, then why not also abandon these core teachings? Some have tried. They live with their boyfriends or girlfriends, pretend they came from nature instead of

God, avoid the Bible and those who believe it, and convince themselves that sin has no consequences. But these same Genesis doubters that characterize our culture have to wear clothes if they want to shop for food without getting arrested. They have to marry to afford the best chances for their children to mature into productive citizens. They understand that engineers make tools to work and yet insist that nature, not an Engineer, made the most exquisite engineering in the world that takes the form of their own human bodies. Those who deny the Bible as the Word of God actually borrow biblical principles to live their lives—for example, assuming that their senses can be trusted, that the world itself is not God, that time flows in one direction, that rights should be defended, and that wrongs should be amended. Shame on us when we deny sin and yet decry injustice. In other words, it's impossible to live real life without core teachings that come from Genesis. Nor is it possible to construct a consistent Christian doctrine without Genesis.

To state it all positively, when we align our minds and lives with the plain and pure statements that Genesis offers, we end up doing things like getting married and doing the work required to become one with our spouse in every aspect of life. We understand the origins and purpose of wearing clothes. We joyfully credit our great Creator with the ingenuity behind all that He has made instead of confusedly offering praise to nature as though hydrogens naturally turn into humans.

With that high view of Genesis, we gain confidence in our ability to understand how all Scripture is interrelated. We interpret its words normally, as we would interpret words in a conversation or even from this very book. By this means (Hebrews 11:3), we learn where we actually came from, why we are here, why we have unfulfilled longings, and how a future garden and its God are set to fulfill those deepest desires. When we elevate Genesis to its well-earned status as the root of all human history, we find in it the very reason why sin has wrecked this world and threatens to wreck our hearts. Genesis tells us what we need salvation from. The pages of Genesis promise a Savior, and the gospels confirm those first words.

Five Unbroken Threads That Wind from Genesis to Revelation

Genesis connects beautifully with Revelation, but we only see it when we treat the words in both of these Bible bookends as meaning what they most plainly say. In other words, even though John's Revelation contains apocalyptic visions, some of the future events of which it speaks and some of the symbols that it uses in those visions to illustrate those events tie right back to Genesis. Of course, certain words lose a bit of their original plainness after translation, but I have found that when I permit the context of the text to narrow a term's range of definitions, the basic idea almost always unfolds. The degree to which I accept or swallow those statements relies not only on my commitment to giving the words their normal meaning but also on my commitment to try my best to identify my own biases and preconceptions, letting the words tell me what they mean instead of me telling the words what I think they should be saying.

I once treated, as has become all too common, Genesis and Revelation as saying anything but what they appear to say. This me-first treatment was like trying to see through cataracts. Any hitches in the transfer of truth from the Bible's pages to my mind occur on my end. The Bible glows and rings with purity even while its divine Author waits on my biases to clear away so I can better grasp what He clearly says.

Just as we doubt the plain meaning of the Bible's earliest chapters, many of us tend to leapfrog the first-person narrative touchpoints found throughout Revelation's prophecies. In the 13 places (more than any book of the Bible) where the text says, "Then I saw," are we to think, "No he didn't...not really"? This attitude robs readers of ultimate answers to our deep longings about our origins and destiny. It also robs us of being able to appreciate the brilliance of the Bible's divine Author, who wove a glorious tapestry of unbroken threads that wind their way from the Bible's start to its finish.

Do You Want to Live Forever?

One such thread traces our longing for long life. Even those with very little knowledge of the Bible want longevity. Within the modern non-Christian tones of the imaginary *Star Wars* universe, Jedi masters leave this world to become everlasting apparitions. What a concept—for Obi-Wan, slain by Darth Vader, to return in disembodied form and tell Luke to go to the Dagobah System or for transcendental Luke to tell Rey that no one's ever really gone. Many movies take up this theme, where their characters live long lives by star travel, cloning, magic, or enlightenment. Even the first civilizations expressed mankind's universal longing for everlasting significance when they memorialized kings with enormous monuments.

Nobody likes goodbyes—let alone the forced goodbye that death brings. That's because, according to Genesis, death intruded into a one-time "very good" creation. Why do the elderly pine for more youthful times as they lament the miseries of bodies breaking down? This universal desire for perpetual youth sharply contrasts with the reality of our brief and fading lives. Animals don't pine for long life. They just take events as they come. Where did this human sensibility come from if not from the reality of a once-good garden where God made humans to fellowship forever with Him? God then shackled Adam, Eve, and all their descendants, including you and me, to the necessary and just consequences of our sin.

This death curse fits our daily reality. It also fits the Bible's other bookend where the text says, "And there shall be no more curse, but the throne of God and of the Lamb shall be in it, and His servants shall serve Him" (Revelation 21:3). How long shall His servants serve Him? Forever. For "whoever believes in Him should not perish but have eternal life" (John 3:15). Genesis sets forth the beginnings of our longing for eternal life, and Revelation reveals God's plan to fulfill that very longing.

Utopia

Utopia describes an ideal society—a place where anyone would love to live. Trees grow tall, and people play in parks. There one finds friendly neighbors, clean air, no crime, plenty of food, freedom to create and invent, encouragement all around, happy animals, serene and bright weather with no natural disasters, people who skip and sing and who assume good motives in others, and children playing in total safety. In utopia, wives smile at the future, husbands cherish their brides, and sons and daughters seek wisdom and hold virtue in high regard. Such a place, such a community, has never existed on Earth. How could it exist with humankind as selfish as we are and amidst a world full of death and thorns? But our idea of utopia must have come from somewhere. No animal seems to have even a foggy notion of utopia. Why do we seek it, and could it ever somehow happen?

This concept constitutes another unbroken thread that began in the beginning. There, "the LORD God planted a garden eastward in Eden, and there He put the man whom He had formed. And out of the ground the LORD God made every tree grow that is pleasant to the sight and good for food" (Genesis 2:8-9). I'm reminded of fictional Rivendell in *The Lord of the Rings*. And yet prophets who came after Moses promised a real return of utopia with no hint of fiction in their words. Isaiah wrote, "For the LORD will comfort Zion, He will comfort all her waste places; He will make her wilderness like Eden, and her desert like the garden of the LORD; joy and gladness will be found in it, thanksgiving and the voice of melody" (Isaiah 51:3).

Threatening animals don't fit utopia either. I recall the 20 hours I got stranded in a remote Colorado mountain village called Cuchara. I was on the phone with my wife, thankful for that single bar of cell signal I found when I stood on my tiptoes at the top of town. As I relayed miseries about how and when I hoped to repair the broken van, a big brown bear lumbered out of the trees toward a dumpster across what passed for the town street. My one thought was that this animal was big enough and had the right equipment to do whatever it wanted with my frail form. Manliness melted. My heart thumped down to my stomach. This was no utopia.

I quickly barricaded myself in the van and locked its doors, hoping for a clear and quick path from there to my rented room beyond the bear and beyond cell service. Soon enough, a villager walked from the bar and in the bear's direction. I took the chance, remembering the old adage that you don't need to be the fastest person to outrun a bear—you just have to not be the slowest.

I found my bed with no harm done in the end. Neither of us suffered damage or even much threat, but teeth and claws have mauled many men and women through the millennia. The vicious application of fangs and talons belong nowhere in utopia. Someday, though, the fearsome "wolf also shall dwell with the lamb. The leopard shall lie down with the young goat. The calf and the young lion and the fatling together, and a little child shall lead them" (Isaiah 11:6). Do you believe that this God-made utopia will really happen one day on a new earth that He will create? If not, why not? Have you made the proper preparations to ensure your presence there?

At the end of the book of Job, God calls Job's attention to an enormous and terrifying amphibious reptilian sea creature called leviathan. The text provides good reason to treat this as a real animal and one much more awesome than a modern crocodile, which does not fit the habits or anatomy of leviathan as described. Perhaps it was a post-Flood version of some creature that we know from a fossilized pre-Flood ancestor that has gone extinct since Job's time. Maybe it was something like the largest theropod known, *Spinosaurus egypticus*, or the only extinct animal with a mysterious extra pair of nostril-like holes in its snout, the 35-foot-long *Deinosuchus riograndensis*. Shockingly, the all-loving God of the Bible says about this terror:

> I will not conceal his limbs, his mighty power, or his graceful proportions. Who can remove his outer coat? Who can approach him with a double bridle? Who can open the doors of his face, with his terrible teeth all around? (Job 41:12-14)

These enormous extinct reptiles could almost swallow you whole. Here the Lord seems to be saying that as terrifying as those terrible teeth are, the One who made this creature carries much more terror for those who would oppose Him. If He would recreate the world and place on it inhabitants who all chose not to oppose but to fear and follow Him, then the need for biological object lessons from toothy monsters would wash away. And indeed, Revelation describes a new heavens and a new earth. "And there was no more sea" (Revelation 21:1), and thus no more habitat for leviathans. No more terrible teeth.

The Lord handcrafted a utopia in the beginning. Worldwide, citizens of all stripes and all times have since longed for a place of peace and comfort. Revelation details the fulfillment of this promise, saying, "And God will wipe away every tear from their eyes, there shall be no more death, nor sorrow, nor crying. There shall be no more pain, for the former things have passed away" (Revelation 21:4). If we embrace Genesis, which sets up the tension felt since God tasked an angel to guard the gates of Eden, then we might as well accept that perfect resolution of this tension as described in Revelation. It tells of glorious and golden roads and of a tree of life that somehow grows on both sides of a river.

Utopia was once real, and God will remake it. Our longing for a utopia seems like another way to express our desire to return to the garden—a place without death or decay, where every citizen is highly esteemed, and where each citizen has personal access to the Son of Man at the very top. Again, Genesis sets up the story of human history, and Revelation describes how the tension will resolve.

The Devil and the Seed

Many of those who reject the Bible seem to accept the devil as real. The prince of darkness seems to enjoy more time in anti-Christian Hollywood movies than he does in pulpits. Whether or not one chooses to believe in the devil, he certainly has a part in the events laid out in the Bible. According to that same Bible, he plays a key role in the lives of billions of people, making him tough to ignore. Is he real? Where did he come from? What can he do? What will happen to him? His story begins in the garden and winds all the way to the Bible's final pages.

The devil shows up in the Garden of Eden to deceive Eve. Genesis 3 refers to him as "the serpent." There, as he does to us today, he first encouraged doubt of God's Word, saying, "Has God indeed said…?" (Genesis 3:1). I'm writing this book to rebuild confidence in God's Word—confidence that the devil's world system steals from us. He encourages us to doubt God's Word. When we do that, we inevitably think wrong, which leads us to live wrong. After we make those mistakes that Scripture calls sin, the devil turns around and blames us for our folly.

He denied God's Word, saying, "You will not surely die" (Genesis 3:4). Eve followed his lead, and then Adam followed hers. God then issued the curses noted in chapter 1 of this book. Amidst the curse pronounced on the serpent—that devil of old—the Lord embedded a promise pregnant with premonition. The Seed of the woman would bruise the serpent's head. The Lord Jesus is that promised Seed, born of a woman (but not of a human husband), and is thus fully God and fully man. Jesus' resurrection defeated death and the devil, whose ultimate judgment is described in Revelation—the final fulfillment of the Seed of the woman who crushes the head of the serpent.

The prophet Ezekiel described the slippery slope down which this same slimy serpent slunk.

> You were the seal of perfection,
> Full of wisdom and perfect in beauty.
> You were in Eden, the garden of God;
> Every precious stone was your covering:
> The sardius, topaz, and diamond,
> Beryl, onyx, and jasper,
> Sapphire, turquoise, and emerald with gold.
> The workmanship of your timbrels and pipes
> Was prepared for you on the day you were created.
> You were the anointed cherub who covers;
> I established you;
> You were on the holy mountain of God;
> You walked back and forth in the midst of fiery stones.
> You were perfect in your ways from the day you were created,
> Till iniquity was found in you. (Ezekiel 28:12-15)

The devil was created. He was perfect until he became vain and proud and sought to elevate himself above God. What iniquity was found in him? "Your heart was lifted up because of your beauty; you corrupted your wisdom for the sake of your splendor" (Ezekiel 28:17). He decided to begin thinking a lot of himself. Although God established him, perhaps he instead insisted that he had established himself, similar to the way we think that natural processes established us from lowly stuff.

The reference to "corrupted [his] wisdom" must mean that he deceived himself. How so? Maybe he began to think that he was not created after all. He would thus assert that he was greater than God.

How descriptive of so many in our culture. Wouldn't a devil feel thrilled that we have replaced the God of creation with the unholy trinity of time, chance, and death? Like the devil, we have convinced ourselves that we deserve to live life our way, not His way. And like the devil, we have ignored and discounted the very words of God. No wonder Jesus spoke this way to self-righteous religionists who deliberately ignored the overwhelming evidence of God through Jesus' many miracles, saying, "You are of your father the devil, and the desires of your father you want to do. He was a murderer from the beginning, and does not stand in the truth, because there is no truth in him. When he speaks a lie, he speaks from his own resources, for he is a liar and the father of it" (John 8:44).

The bit in Genesis 3:15 about the woman's Seed refers to the Savior in particular. After all, the father plants seed into a mother's womb. But Jesus would be different. He would come straight from the mother—a hint from Genesis 3 later made explicit in Isaiah 7 and later yet fulfilled through Mary, the mother of Jesus. And yes, we desperately need a qualified rescuer to save us from the devil's lair of lies and transport us, for we have no power to transport ourselves, to the Creator's new kingdom of light and life!

The Lord God reminded His key followers of this promised Seed in the years leading up to His incarnation. He told Abraham, "In your seed all the nations of the earth shall be blessed" (Genesis 22:18). Paul affirmed this when he commented on Genesis 22, saying, "Now to Abraham and his Seed were the promises made. He does not say, 'And to seeds,' as of many, but as of one, 'And to your Seed,' who is Christ" (Galatians 3:16). Here, Paul treated not just the words of Genesis as accurate, but he bases his whole argument on the accuracy of just one letter! Now that's a high view of Genesis. Why should our view of Genesis be any lower than Paul's?

The Lord had promised to Abraham that the Seed would come through Isaac (Genesis 21:12). Then He specified the tribe of Judah (Genesis 49:10). Isaiah foretold that the Seed would belong to the Lord and live forever (Isaiah 53:10). God promised that His Seed would descend from Israel's greatest king, David (2 Samuel 7:12). This is why Scripture contains genealogies in Matthew 1 and Luke 3 that tie David to Jesus. "Remember that Jesus Christ, of the seed of David, was raised from the dead according to my gospel" (2 Timothy 2:8).

The Lord Jesus' reference as the "Seed" does not appear in Revelation, but *He* sure does. The same serpent who bruised the Lord's heel on the cross gets his final showdown against this Jesus whom sinners unjustly planted into the earth. Only He sprouted anew from that grave in order to bring many to Himself.

Lest there be any doubt about the identity of the serpent in the garden, John referred twice to "that serpent of old,…the Devil and Satan" (Revelation 12:9, 20:2). This wicked thread winds his way from Genesis into the hearts of those who wish to elevate themselves over God, all the way to the end of this age and even to the last foretold age. Along the way, he has inadvertently accomplished God the Creator's plans and purposes throughout history. For example, the devil inspired his human followers to put Jesus to death, thinking that would be the end. Nope. That was just the end of sin's tyranny over the hearts of all who would trust and follow Jesus, then and now.

Ezekiel foretold the devil's fate, saying, "You have become a horror, and shall be no more forever" (Ezekiel 28:19). Revelation tells more details. "The devil, who deceived them, was cast into the lake of fire and brimstone where the beast and the false prophet are. And they will be tormented day and night forever and ever" (Revelation 20:10). Who cast the devil into the lake? Jesus. This portrays the part where the Seed finally and completely crushes the serpent's head. The problems revealed in Genesis 3 will find their everlasting solutions at the end of history. At that time, a God-made garden-city with a river running through it upstages the first garden. Can we accept John's Revelation and yet set Genesis aside? They stand together as bridge abutments for the whole Bible. And the Lord Jesus is the whole bridge.

Mended Relationships

Genesis records the first broken relationship and its restoration. After Adam and Eve did precisely what God warned them not to do, they immediately knew that they had severed pure relationships with their God. Then, to add to their shame, they hid from God instead of attempting to reconnect with Him. But God, in his love and kindness, sought them. This same plot plays out in the life of every man, woman, and child.

Oh, for a way back! We long—whether we suppress it or not—for that inner peace that comes from a relationship restored through forgiveness. What other way is there? We cannot go back and undo the harm we caused or unthink our evil thoughts. Just as God made a way for the world's first couple, He provides for any and all who will come to Him. "'Come now, and let us reason together,' says the LORD, 'Though your sins are like scarlet, they shall be as white as snow; though they are red like crimson, they shall be as wool'" (Isaiah 1:18). But how? "By the hearing of faith" (Galatians 3:2). Thus, Isaiah echoed the thread that winds from Genesis to Revelation that our relationship with God needs mending. God clothes. God mends. Our Lord longs to restore.

This leads us to Revelation where this thread finds its end. "The Lamb of God who takes away the sin of the world" (John 1:29) removed sin's penalty. This very Lamb, to whom all those Old Testament sacrifices pointed, offers to apply His own shed blood to cover even the sins of we who shake our fists at Him. But only those who humble themselves—lower those fists—actually receive that applied blood to enter an everlasting reconciliation. These individuals "taste and see that the LORD is good" (Psalm 34:8). Those saved from sins find the ultimate relationship restored.

After He mends our hearts, He will mend the whole universe. He promises to set up a new and holy city that lasts forever. No wonder those who disbelieve Genesis creation have a hard time sorting through Revelation. After all, a God who didn't actually create the first universe the way He said He did can't be expected to create another universe the way He promises He will. Only by taking Scripture's words at face value do they unspool the unbroken thread of mending relationships and building a special place for those relationships to thrive forever (John 14:2).

In that place, He will establish those whom He makes holy and who gain endless days to express gratitude for forgiveness and new life. "And there shall be no more curse, but the throne of God and of the Lamb shall be in it, and His servants shall serve Him" (Revelation 22:3). How glorious that this King of creation will not only mend broken relationships but nourish and cherish each of them forevermore!

A real Adam and Eve broke their real relationship with a real God. So do we when we sin. We have a chance to find forgiveness. We can seize it by submitting to Him. The Bible's main narrative of humankind reconnecting with God finds its foundation in a historical Adam who severed that connection in the first place. Praise the Lord that He made a way back through His Son, the last Adam, by whose mortal wounds we are healed!

A Name

The final of our five unbroken threads that wind from Genesis to Revelation is a name. In Genesis 1, God names specific features of His creation. The first feature to receive a name was the daily light/dark cycle of Earth, wherein "God called the light Day and the darkness He called Night" (Genesis 1:5). Naming things seems important to the Lord. He even expressed intrigue of what names others might come up with. "Out of the ground the LORD God formed every beast of the field and every bird of the air, and brought them to Adam to see what he would call them" (Genesis 2:19). Your name becomes the symbol of your identity just as God's name symbolizes who He is. This penchant for giving and receiving names—and for names to encapsulate identities—winds its way from Genesis through our present world and on to Revelation.

Abram humbled himself before the Lord, seeking no name for himself. Therefore, God exalted him. "Humble yourselves in the sight of the Lord, and He will lift you up" (James 4:10) remains as helpful a principle today as it was back then. God told Abram, "And I will make of thee a great nation, and I will bless thee, and make thy *name* great; and thou shalt be a blessing" (Genesis 12:2, emphasis added).

This blessing contrasts with a very different message God delivered in Genesis 11. There, a throng of narcissists tried to thwart God's plans as they built the Tower of Babel. They said, "Let us build us a city and a tower, whose top may reach unto heaven; and let us make us a *name*, lest we be scattered abroad upon the face of the whole earth" (Genesis

11:4, emphasis added). Does anyone today know any of their names? Nobody does. So much for their attempt to glorify themselves. But few fail to learn of Adam or Abraham.

King David enjoys a famous name, too. Like Abraham, he did not earn his own name. Rather, his God granted it to him. David has the reputation of being a man after God's own heart. Though certainly sinful, he was characterized by lifelong efforts to glorify God instead of himself. God honored him and blessed him, even promising to bring about the Savior—that Genesis 3 Seed—through David's lineage. God spoke to David about names through the prophet Nathan, saying:

> "I took you from the sheepfold, from following the sheep, to be ruler over My people, over Israel. And I have been with you wherever you have gone, and have cut off all your enemies from before you, and have made you a great name, like the name of the great men who are on the earth....
>
> "When your days are fulfilled and you rest with your fathers, I will set up your seed after you, who will come from your body, and I will establish his kingdom. He shall build a house for My name, and I will establish the throne of his kingdom forever." (2 Samuel 7:8-9, 12-13)

What higher name could anyone bear than one of those listed in the genealogy of the Lord Jesus Christ Himself? (See Luke 3:31.)

This trend winds its way to you and me. We each long for recognition. That's why as children (or as middle-aged people who retain childish ways) we make fun of peers—to elevate ourselves over others. That's why some scientists labor long hours in their labs—to possess the most renowned name in their field. Name elevation motivates some high-powered businesspeople to do whatever they can so long as it elevates their name to the status of "rich and successful." Too late do some of us realize that it's much better to "let another man praise you, and not your own mouth; a stranger, and not your own lips" (Proverb 27:2).

Stepping off the treadmill of self-exaltation means relinquishing control over our own value of personhood. We have to let someone else decide what we're worth. That's a terrifying prospect when the One doing the deciding is a holy Creator and we are unholy at heart. This is why the God of the Bible taught His disciples:

> "If anyone desires to come after Me, let him deny himself, and take up his cross, and follow Me. For whoever desires to save his life will lose it, but whoever loses his life for My sake will find it. For what profit is it to a man if he gains the whole world, and loses his own soul? Or what will a man give in exchange for his soul? For the Son of Man will come in the glory of His Father with His angels, and then He will reward each according to his works." (Matthew 16:24-27)

According to Revelation, at that time each man and woman "who overcomes" will receive a new name.

> "He who has an ear, let him hear what the Spirit says to the churches. To him who overcomes I will give some of the hidden manna to eat. And I will give him a white stone, and on the stone a new name written which no one knows except him who receives it." (Revelation 2:17)

Thus, our universal longing for a long-lasting, good name has its origins in Genesis, resonates with what our own hearts admit about ourselves, and finds its fulfillment in the Bible's other bookend.

What emerged from my transformation from taking Genesis lightly to taking it seriously? What I've been trying to convey in this discussion is that only when I treated the parts, places, and people in Genesis as real gardens, real trees, real pre-Flood lands, and a real Adam, Eve, and devil did I discover that those same elements play precise roles in this history in which even I get to play a real part. To erase those elements by calling them symbols instead of realities erases them from throughout the Bible. It erases them from Revelation, which diminishes our hope. On the other hand, when I let those elements be what Scripture says, my hope in taking part in a real future with the true Savior and His new Earth found more sure footing. And who can really live without hope?

What the Prophets Thought of Genesis

In my journey toward rediscovering the reliability of Genesis, I began to sift through the books of the Old Testament prophets to see how they treated it. I discovered a unanimous approach. They all took for granted a historical Genesis. They even quoted specific names of people and places as history, not as poetry or allegory. What gave me the right to treat early Genesis differently than the prophets? Does conventional science give a strong enough reason to call into question not only Genesis itself but those whose very words tie back to Genesis? Chapter 3 in this book argued that science affords us no such basis. I found instead that plenty of science adds reasons to support the prophets' view of Genesis. But the current task is to let key passages show how the prophets treated Genesis 1–11 as history.

Was Adam a real person? Naturalistic scientists, wedded as they are to a worldview that denies miracles and to a narrative of human ancestry from ape-like creatures, must insist that Adam was a myth. However, 1 Chronicles 1:1 names Adam as the first man in its opening genealogy. If Adam was not a man, then how and when did his listed descendants morph from raw fantasy to real fathers?

Jewish tradition holds that Ezra the priest wrote Chronicles, which is one book in the Hebrew Bible and two in the Christian Bible. The book bearing his name notes that "Ezra had prepared his heart to seek the Law of the LORD, and to do it, and to teach statutes and ordinances in Israel" (Ezra 7:10). Tradition tells that Ezra committed the Scriptures to memory, even as a few lovers of the Bible today have done. This great scribe was traditionally responsible for the books Ezra, Nehemiah, and both books we today call Chronicles. If he was wrong about Adam in one of his books, then why should we trust his other books? Whoever wishes to refute him bears the burden of explaining why all these books no longer deserve to be considered Scripture when they've been treated as such since the end of Jewish Babylonian captivity in 538 BC.

Job wrote, "If I have covered my transgressions as Adam, by hiding my iniquity in my bosom..." (Job 36:33). Certain Christian scholars tell churchgoers not to embarrass Christianity by teaching something as scientifically untenable as Adam. Do they want us to reject Job, too, since he referenced Adam as real? If the books of Job, Nehemiah, and the rest never did belong in the Bible, then why didn't prophets and other God-fearers of that time reject the mistaken texts as belonging to Scripture right then and there?

Where did Adam and Eve live in the beginning, according to Genesis? The garden, of course. The prophets also regarded the garden as a real place. "You were in Eden, the garden of God" (Ezekiel 28:13). "For the LORD will comfort Zion, He will comfort all her waste places; He will make her wilderness like Eden, and her desert like the garden of the LORD" (Isaiah 51:3). "The land is like the Garden of Eden before them, and behind them a desolate wilderness; Surely nothing shall escape them" (Joel 2:3). Some might say that Eden was not real but merely *represents* an ideal place. But nowhere does the Bible itself give such a meaning to the name. Instead, it treats the name as though it refers to an *actual*, historical, ideal place. Do we reject Isaiah and Joel, too, since they use the name to refer to an actual place?

Removal from Scripture of the Bible's books that name Adam or Eden erases 26% of the Old Testament. If one were to identify where Scripture quotes or references other Scripture, one would soon find that all the books, directly or indirectly, tie to one another. If Genesis goes, then why not jettison all of it?

Old Testament books that refer to Adam, Eden, and Noah as historical

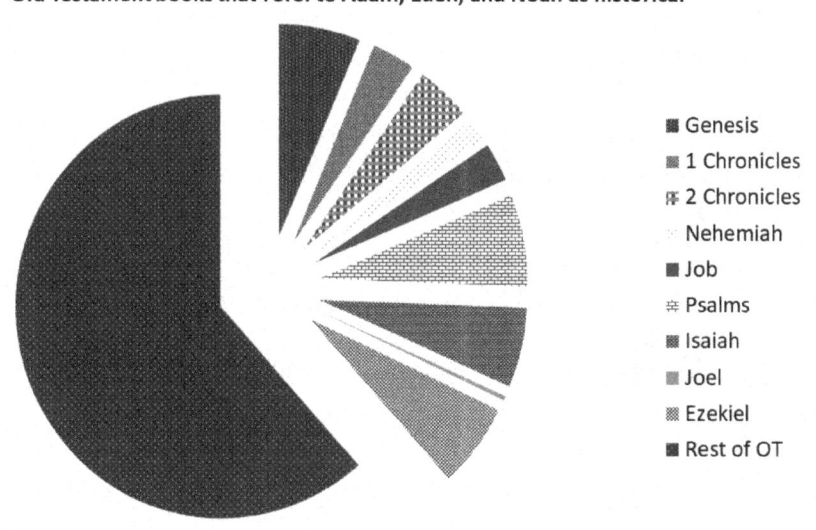

- Genesis
- 1 Chronicles
- 2 Chronicles
- Nehemiah
- Job
- Psalms
- Isaiah
- Joel
- Ezekiel
- Rest of OT

That's exactly what many of our friends and family have done. This book is my attempt to explain why our culture should turn its scorn for the first book of God's Word into an awe over how its accuracy and elegance point to a divine Author who is worth our wholehearted admiration.

The same could be said of the prophets' treatment of the much-maligned Noah and his Flood. "'Even if these three men, Noah, Daniel, and Job, were in it [an unfaithful land], they would deliver only themselves by their righteousness,' says the Lord God" (Ezekiel 14:14). Noah the man, not the figment. Isaiah and 1 Chronicles also name him. Psalm 29:10 names Noah's Flood using the distinct Hebrew word *mabbul*. The prophets treated the events described in Genesis as though they really happened. And I started to wonder if I really had a strong enough basis to say they were all wrong. Jesus said it best when He said, "For if you believed Moses, you would believe Me; for he wrote about Me. But if you do not believe his writings, how will you believe My words?" (John 5:46-47). He was calling me to believe all His words, and He was patient enough to give the time to get all my questions answered before I decided to say "yes."

What the Apostles Thought of Genesis

What did the prophets and the apostles have in common? They knew the Lord, of course. The Lord also used them to write Scripture. The apostles were personally tutored by the Lord Jesus. The apostle Peter urged his readers to hearken to these words over and above the words of sinful men who he predicted would deny the prophets' writings, even as has become the attitude today. He wrote:

> Be mindful of the words which were spoken before by the holy prophets, and of the commandment of us, the apostles of the Lord and Savior, knowing this first: that scoffers will come in the last days, walking according to their own lusts, and saying, "Where is the promise of His coming? For since the fathers fell asleep, all things continue as they were from the beginning of creation." For this they willfully forget: that by the word of God the heavens were of old, and the earth standing out of water and in the water, by which the world

that then existed perished, being flooded with water.
(2 Peter 3:2-7)

Peter treated the Genesis 1 creation account as real, too. His whole argument hinges on Genesis as a precise and trustworthy recounting of our beginnings, even going so far as to reference the world's watery origin, when "the Spirit of God hovered over the surface of the deep [waters]" (Genesis 1:1). Peter treated the Genesis Flood as a real and global destruction. This apostle goes one better by predicting that the future would see folks intent on scoffing at these two great miraculous yet historical events described in Genesis: the creation and the Flood. Who can reasonably deny that we live in that foretold world even now?

Paul in Romans, 1 Corinthians, and 1 Timothy refers to Adam by name. So does Jude, the brother of Jesus, who also calls his readers' attention to other incidents relayed in Genesis. Indeed, Paul wrote half the New Testament and pins the great doctrine of original sin onto our physical forefather. That first Adam is the reason why God supplied His Son to be the last Adam (1 Corinthians 15:45). Delete the first Adam and we might as well delete the last half of the New Testament while we're at it. There goes the Good News!

New Testament books that name Adam or Noah as real people

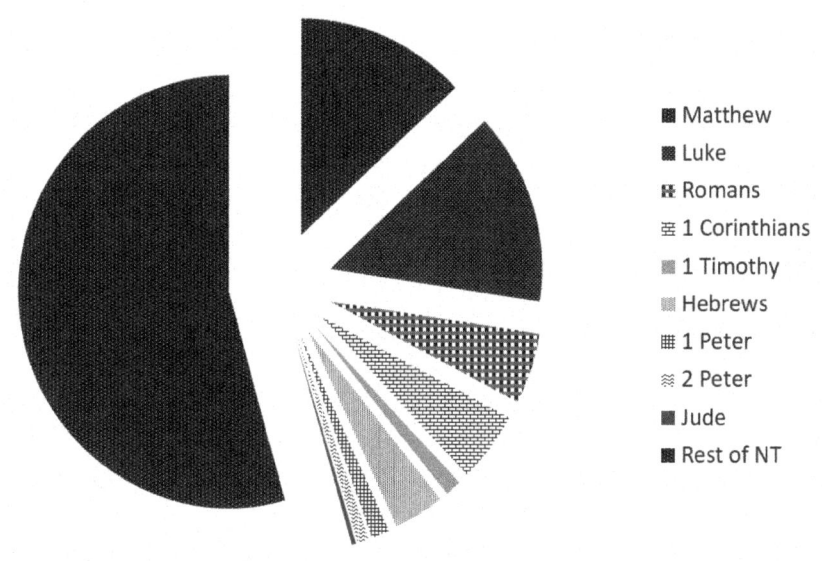

- Matthew
- Luke
- Romans
- 1 Corinthians
- 1 Timothy
- Hebrews
- 1 Peter
- 2 Peter
- Jude
- Rest of NT

Fortunately, science actually confirms Adam in several ways as already mentioned in chapter 3. The point here is that Peter and Paul believed in biblical creation. That is to say, they took Genesis at face value. Peter went so far as to encourage Christ-followers to remind themselves of the words of the prophets, and of course that includes Genesis.

The author of Hebrews wrote, "By faith Noah, being divinely warned of things not yet seen, moved with godly fear, prepared an ark for the saving of his household, by which he condemned the world and became heir of the righteousness which is according to faith" (Hebrews 11:7). Similarly we read, "Once the Divine longsuffering waited in the days of Noah, while the ark was being prepared, in which a few, that is, eight souls, were saved through water" (1 Peter 3:20). Scripture presents Noah as a real man.

The apostles left other evidence of their high view of early Genesis. For example, consider Romans 1:20: "For since the creation of the world His invisible attributes are clearly seen, being understood by the things that are made, even His eternal power and Godhead, so that they are without excuse." How long have folks been able to understand and clearly see His attributes? For millions of years? No—right from the beginning during the creation week when God handcrafted the first human couple.

Hebrews assumes the same recent creation. "Not that He should offer Himself often, as the high priest enters the Most Holy Place every year with blood of another—He then would have had to suffer often since the foundation of the world; but now, once at the end of the ages, He has appeared to put away sin by the sacrifice of Himself" (Hebrews 9:26-27). If the Lord would have had to suffer to fulfill His priestly role "since the foundation of the world," then there were sinners who needed their sin "put away" since the very beginning. According to Hebrews 9, sin started not since the foundation of modern man who emerged from a messy pile of ape-like predecessors long *after* the world began but *since the foundation* of the whole world as described in Genesis 1–3. The apostles wrote or approved the New Testament books. They treated Genesis as history. Should any Bible believer treat it any less?

How Jesus Treated Genesis

The Lord Jesus had the same high view of biblical creation. He said this about coming judgment: "For in those days there will be tribulation, such as has not been since the beginning of the creation which God created until this time, nor ever shall be. And unless the Lord had shortened those days, no flesh would be saved" (Mark 13:19-20). This verse assumes that human "flesh" existed "since the beginning of the creation." If so, then Jesus here refutes the mainstream notion of a beginning billions of years before humans. The Lord Jesus Himself also referred to Noah and his Flood as a real person and a true event (Luke 17:27), as though Genesis needed no amendments.

The *Henry Morris Study Bible* notes a similar argument about Mark 10:6-8. There, Jesus quoted from Genesis 1 and 2, saying, "But from the beginning of the creation, God 'made them male and female.' 'For this reason a man shall leave his father and mother and be joined to his wife, and the two shall become one flesh'; so then they are no longer two, but one flesh." Dr. Morris wrote:

> Mark's account adds the important additional information that the Genesis record from which he is quoting referred to "the beginning of the creation." That is, God made Adam and Eve (Genesis 1:26-27), right at the very beginning—not some 4.5 billion years after the beginning, as modern evolutionists and theistic uniformitarians would have us believe. See also Mark 13:19. There is no room whatever in Scripture for the geological ages, and Christians who compromise on this issue for the sake of academic acceptance are undermining God's Word.[2]

How could God's only Son possibly err? After all, He was "appointed heir of all things, through whom also He made the worlds; who being the brightness of His glory and the express image of His person, and upholding all things by the word of His power, when He had by Himself purged our sins, sat down at the right hand of the Majesty on high" (Hebrews 1:2-3).

In sum, the prophets, apostles, and the Lord Jesus Himself treat

Genesis as history. These men qualify as reliable witnesses plus God's spokesmen—and one of them as God Himself. Now, this made me curious. What other evidence did I need before I would believe in all of Scripture? The sciences, societies, spiritual teachings, and the Scriptures themselves all pointed the same direction. The evidence mounted in my mind until it tilted that side of the scale.

But could I now call the support for biblical creation "beyond reasonable doubt"? I wanted first to know if believing in Genesis was in any way important to understanding fundamental truths like what I know about me, about the gospel, and about God. That's what we will consider next.

6
GENESIS AND TRUTH

"Who is it who decides what is important and what is unimportant? Who is to tell us what to pick out and what to leave behind? Where is our standard? Where is our authority? Now as I want to show you there is no such distinction drawn in the Scripture at all."[1]
— D. M. Martyn Lloyd-Jones, 1963

Where to Draw the Line?

Attend one of probably most Old Testament survey courses and you'll learn why Bible scholars have a difficult time with Genesis 1–11. They often draw a line before Abraham, whose story begins in chapter 12. After all, how can they take verses that say "God created" animals, plants, Adam, or Eve seriously when so much science has proven that man, animals, and plants all evolved from the same original creature?

According to mainstream scientists—who after all wear white lab coats, use fancy gizmos, and speak in strange tongues—humans emerged from natural processes and not supernatural ones. The decision to take the side of evolution over what the text says forces those Bible readers to try and decide where errors end and facts begin. I felt that if I could find a pattern within Scripture that justified fact/fiction borders, I would feel more confident about why I didn't need to take Genesis so straightforwardly.

I never found that pattern.

Genesis rolls out events smoothly right between chapters 11 and 12. When other passages speak of creation, biblical authors attribute it to miracle and credit God. Their recounting of God creating stars, animals, or man all reads just as historical as their recounting of Abraham, Moses, or David.

For example, what about Isaiah 37:16? "O LORD of hosts, God of Israel, the One who dwells between the cherubim, You are God, You alone, of all the kingdoms of the earth. You have made heaven and earth." If evolutionary origins really happened, then on what basis would God accept credit for what nature actually achieved in making the worlds?

And then there's John 1:3. "All things were made through Him, and without Him nothing was made that was made." Does "all things" include stars? If so, mainstream scientists had assured me that natural processes that mostly involve gravity made stars, not God. And yet I could find no line or even fuzzy zone in the Bible that would allow me to keep my creation-by-nature stance. "You are worthy, O Lord, to receive glory and honor and power; for You created all things, and by Your will they exist and were created" (Revelation 4:11). Genesis through Revelation take the same tone in giving God all the credit for His supernatural efforts, not nature's, in creation.

Keep on attending that Old Testament survey class and you might hear how archaeology supposedly contradicts early Genesis. Many teach that certain discoveries demand that the words must not mean what they appear to say. A young friend of mine happened to attend an Old Testament survey class when I wrote this. He told me of his disappointment (as one who believes Genesis) upon hearing from his Baptist seminary professor that archaeology shows no evidence for a worldwide flood. The solution? Instead of questioning whether or not archaeology is even the most appropriate discipline to reveal evidence for Noah's Flood, his professor's stance forces him to recast the words "all the high hills under the whole heaven were covered" (Genesis 7:19) to mean "some of the low areas under part of the heavens were covered." What use are words that can mean whatever we what them to mean?

Similar arguments come from radiocarbon dating results that show human and other artifacts deposited tens of thousands of carbon years ago. I say carbon years because my studies have shown that carbon years often diverge from calendar years. But my friend's Old Testament professor and thousands like him do not generally investigate the assumptions that enter the radiocarbon dating process. They too often take at face value results with their baked-in conventional biases. They, as I once did, assume therefore that the Genesis chronologies cannot convey trustworthy time stamps.

But this conclusion concocts new tensions. For example, the chronologies concern those patriarchs whose line of descent ultimately proves the Lord Jesus Christ's humanity and His legitimacy to be the promised Savior of all humankind. How can we peel off their ages but keep their names if the text treats both aspects with equal dignity? And if we cannot trust the number of years that ancestor X lived, then what about the actual ancestor? This exercise erodes confidence in Jesus' legitimacy to inherit His everlasting kingdom, for He must be both the Son of God and the Son of Man—*the* Seed of the woman—to take the throne.

I think we too often rely on science's ever-changing pronouncements about the past to draw ever-changing fact/fiction distinctions in the Bible. It seems that whoever wants to know what it really means would have to get a Ph.D. in the appropriate science. This tendency threatens the doctrine of biblical clarity. Scripture would then lose its ability to speak to its target audience—anyone with ears to hear (Matthew 13:9).

Have conventional scientists become sort of like ancient priests—here to guard which parts to believe and which to ignore? "All Scripture is given by inspiration of God, and is profitable for doctrine, for reproof, for correction, for instruction in righteousness, that the man of God may be complete, thoroughly equipped for every good work" (2 Timothy 3:16-17). The Bible is supposed to instruct and equip "the man of God," not the "the scientifically trained man of God." I began to recognize that since the historical narrative texts within Scripture all sound the same, ever-shifting decisions of where to stop believing it come from outside the Bible. In other words, they come from the minds of fallen sinners like me.

God was wooing me to believe it all. Would I trust the words of fallible humans even if they wear lab coats or the Word of God that opposes their claims? Is all of Genesis true? The text itself seems to say, "yes." Could I live an intellectually truthful life if I believe some Genesis stories as history but refuse to believe other Genesis stories even though they take the same historical format and tone? I decided that I could not live a double life—a lie. I had to know more. I turned next to considering the gospel truth.

Genesis and the Gospel

Romans Road Runs Through Genesis Way

A man once told me he was a "New Testament Christian." By this he meant that he did not believe in Genesis as history, but he did believe in Jesus. I asked him if he could think of any New Testament doctrine not based in Genesis. He couldn't, although to be fair he was unfamiliar with the word "doctrine." It simply means teaching. The gospel summarizes the core teaching of Christianity. All the core concepts in the gospel hinge on Genesis. The apostle Paul pointed the gospel back to Genesis by referencing "Scriptures" in 1 Corinthians 15:3-8, saying:

> For I delivered to you first of all that which I also received: that Christ died for our sins according to the Scriptures, and that He was buried, and that He rose again the third day according to the Scriptures, and that He was seen by Cephas, then by the twelve. After that He was seen by over five hundred brethren at once, of whom the greater part remain to the present, but some have fallen asleep. After that He was seen by James, then by all the apostles. Then last of all He was seen by me also, as by one born out of due time.

This specifies that the Lord Jesus Christ fulfilled Scripture's prophecies and expectations of a perfect man/God who would die for our sins (Isaiah 53:5) and then rise again after three days (Matthew 12:40).

Can you think of any key Christian doctrine that does not find its roots in Genesis?

Genesis introduces right and wrong, God's law, and His expectations. Genesis establishes human failure and our utter lack of ability to set things right, either in our lives or the lives of those we damage whether by accident (Genesis 12:17) or intent (Genesis 4:8). Genesis demonstrates God's salvation by grace. When all but eight humans completely rebelled against God, He did not erase all mankind. He saved Noah, Shem, Ham, Japheth, and each of their wives because "Noah found grace in the eyes of the LORD" (Genesis 6:8).

The Romans Road to salvation highlights four verses from Romans that summarize what any guilty sinner needs to know and do to be saved from judgment. Each stop along the road reveals an essential link that securely anchors the gospel back to Genesis.

At the first stop along the Romans Road, we learn that "all have sinned and fall short of the glory of God" (Romans 3:23). But what is sin? It means breaking God's holy law. God gave the first law to Adam and Eve in the garden, saying, "But of the tree of the knowledge of good and evil you shall not eat" (Genesis 2:17).

The second stop on the Romans Road says, "For the wages of sin is death, but the gift of God is eternal life in Christ Jesus our Lord" (Romans 6:23). Death refers to separation. This can refer to the body separating from the soul or a person being permanently separated from God—a second death (Revelation 21:8). Genesis 3 tells where death came from. When Adam and Eve disobeyed God, justice demanded the death penalty. When evaluating and comparing their act of disobedience against the Ten Commandments given in Exodus 20, by my count their eating of the forbidden fruit broke at least commandment numbers one, five, eight, nine, and ten.

1. "You shall have no other gods before Me" (20:3). They made themselves like gods who assumed the wisdom and power to do what's best despite God's warning to the contrary.
5. "Honor your father and your mother, that your days may be long upon the land" (20:12). They dishonored their Father by dismissing His command.
8. "You shall not steal" (20:15). They stole the fruit.
9. "You shall not bear false witness against your neighbor" (20:16). Eve shifted blame to the serpent, and Adam shifted blame to his wife. They lied to try and cover their guilt.

10. "You shall not covet…anything that is your neighbor's" (20:17). They coveted their neighbor's (God's) fruit and His wisdom.

Adam and Eve thus introduced sin into the cosmos. God's earlier warning about dying as the big consequence of sin became all too real for them and for every human since.

According to evolutionary history, mankind did not descend from Adam and Eve but evolved from animal ancestors that had been living and dying for billions of years. Those who believe this version of history, as I once did, see no Adam in their past and understandably struggle to understand the very concept of sin. Now it's up to Christ-followers to bring to friends, family members, and neighbors a full confidence in the history of the world that God supplies in the book of beginnings.

The third stop on the Romans Road brings some good news. "But God demonstrates His own love toward us, in that while we were still sinners, Christ died for us" (Romans 5:8). This says that although our crimes against God earn death, God the Father sentenced His own Son in our place. Genesis 3:15 promised help through a descendant of Eve, and that help came with Christ's virgin birth, as clearly foretold in Isaiah 7:14 five centuries before its fulfilment: "Therefore the Lord Himself will give you a sign: Behold, the virgin shall conceive and bear a Son, and shall call His name Immanuel." His life, death, and resurrection provided a way to escape everlasting death, just as the Lord long ago provided an ark as the escape from dying in Noah's Flood. Only eight people believed God's message of coming judgment and entered the Ark to be saved.

I pray that the "New Testament Christian" I met notices how the New Testament's gospel relies on Genesis history at every turn. The New Testament even treats the Flood, at which modern conventionalists scoff, as real. "[God] did not spare the ancient world, but saved Noah, one of eight people, a preacher of righteousness, bringing in the flood on the world of the ungodly" (2 Peter 2:5). Matthew, Luke, Hebrews, and Peter all refer to Noah as a real person, as I noted in chapter 3 (Matthew 24:37-38; Luke 17:26-27; Hebrews 11:7; 1 Peter 3:20; 2 Peter 2:5). Plus, Jesus descended from Adam through Noah, according to Luke 3:23-38. If Jesus did not descend from Adam (as revealed in

Genesis), then He would not be our kinsman, or relative. He, being one with the Father, became one of us so that He could fulfill Scripture's expectations of a kinsman-redeemer.

> For both He who sanctifies and those who are being sanctified are all of one, for which reason He is not ashamed to call them brethren, saying: "I will declare Your name to My brethren; In the midst of the assembly I will sing praise to You." And again: "I will put My trust in Him." And again: "Here am I and the children whom God has given Me." Inasmuch then as the children have partaken of flesh and blood, He Himself likewise shared in the same, that through death He might destroy him who had the power of death, that is, the devil, and release those who through fear of death were all their lifetime subject to bondage. For indeed He does not give aid to angels, but He does give aid to the seed of Abraham. Therefore, in all things He had to be made like His brethren, that He might be a merciful and faithful High Priest in things pertaining to God, to make propitiation for the sins of the people. For in that He Himself has suffered, being tempted, He is able to aid those who are tempted. (Hebrews 2:11-18)

Only a human can pay the price for human sin, and only the Lord can live a sinless life. Without His history linked to Adam, how could Christ's payment apply to Adam's descendants like you and me?

He paid our death penalty and defeated death with His resurrection. This way, even though our sins have earned eternal separation from God, the Lord Jesus can now commute to zero this death penalty and save any and all who trust in Him. This brings up the last stop on the Romans Road: how to return to the God who made us and sacrificed His only Son to bring us back to Him.

> That if you confess with your mouth the Lord Jesus and believe in your heart that God has raised Him from the dead, you will be saved. For with the heart one believes unto righteousness, and with the mouth confession is made unto salvation. (Romans 10:9-10)

God initiated this whole doctrine of salvation by grace through faith (Ephesians 2:8) with Abram. Genesis says that Abram "believed in the LORD, and He accounted it to him for righteousness" (Genesis 15:6). Later renamed Abraham, he became the physical father of the nation from which the promised Messiah would come. He became the spiritual father of all who would trust in the Lord. God even encourages His nation Israel to follow Abraham's example of faith in the Lord and to believe what He says. "Look to Abraham your father, and to Sarah who bore you; for I called him alone, and blessed him and increased him" (Isaiah 51:2). Even the New Testament clarifies, "Therefore know that only those who are of faith are sons of Abraham" (Galatians 3:7).

So, every element of the gospel—God's glory, God's law, man's sin, a death penalty, a way of escape, and taking God at His Word—begins in Genesis. I can think of no reason why the "New Testament Christian" I met shouldn't just become a "whole Bible Christian."

Does Recent Creation Hinder the Gospel?

If those who believe the gospel follow this line of thinking to its logical conclusion, then we end up having a hard time rejecting its Genesis foundation. Does that have to include unscientific-sounding parts of Genesis like its record of a recent creation in just six days? I presented solid science that actually supports this view in chapter 3 of this book. Most believers with whom I interact on this subject are unfamiliar with this science. It can really help those who wish they could believe in Genesis but feel intimidated by "science." Many have decided that God created over millions of years, not six days, or that God allowed nature to evolve over millions of years, not by the power of His command. These Christians—probably most Christians—decry Genesis as unscientific. They oppose recent creation and argue that insisting that the world is only thousands of years old *hinders* the gospel by adding an unreasonable barrier to belief. But recent creation actually supports the gospel in two key ways.

The first way that recent creation supports, not hinders, the gospel has to do with the Bible in general. When Christians believe in a recent creation, they show complete confidence in the accuracy and authority of all of God's Word, including its history. Telling unbelievers that they do not have to believe certain miraculous events described in Scripture—like creation, the global Flood, or God parting the Red Sea—but they do have to believe in the miracle of Christ's resurrection sends a confusing message. On what basis should a person select which events—miracles or not—actually happened? Since God knows everything, we can believe all the miracles and other historical occurrences He recorded in Scripture. After all, the same New Testament authors who saw and described the resurrected Lord also referred to recent creation (Hebrews 9: 26), Adam (1 Corinthians 15:22), the Flood (2 Peter 3:5-6), and the Red Sea parting (1 Corinthians 10:1-2). A whole Bible Christian, not a parts-of-the-Bible Christian, can express confidence that anyone can understand the Bible's answers to our most pressing questions of origins, purpose, and destiny.

Another way that recent creation supports the gospel is by supplying its foundation. The only reason our salvation would require Jesus' death is if our sins really earn eternal death. And they do. "Therefore, just as through one man sin entered the world, and death through sin, and thus death spread to all men, because all sinned..." (Romans 5:12). Sin and death affect more than mankind, "for we know that the whole creation groans and labors with birth pangs together until now" (Romans 8:22).

Now, if the whole creation is billions of years old, then according to the evolutionary interpretation of the billions of fossils encrusted in planet Earth, uncounted numbers of animals were dying long before the Genesis 3:19 death penalty ("for dust you are, and to dust you shall return") for Adam's sin. Genesis 1:31 says the world was "very good" before the curse of death. It must not have been filling up with dead animals millions of years before Adam arrived. So, where did those fossil come from? Since Noah's Flood involved worldwide watery catastrophe and fossils the world over were all buried in catastrophic water deposits, then it makes sense to say fossils formed during Noah's recent Flood, which happened long *after* the curse. The very foundation for Jesus paying our death penalty relies on a creation that God cursed *after* Adam's sin.

Can a person believe in Christ and still believe in evolution or millions of years? Yes! I was one of them. But those who grow in Christ should always be "bringing every thought into captivity to the obedience of Christ" (2 Corinthians 10:5). If His Word tells us the very good and true news of salvation from sin's penalty by God's grace, conveyed to individuals through our trust in Him, then this same Word ought also to convey the true history of the world. Recent creation affirms "the whole counsel of God" (Acts 20:27) and establishes that death follows sin—the very problem that the gospel solves.

Some have tried to circumnavigate this second argument. One approach involves insisting that the sin in the garden brought on the death of mankind only, not of animals. I received a letter to this effect once. This Christian insisted that rock layers could only have formed over millions of years. In reality, rock layers dozens of feet thick can form fast, and virtually all of Earth's sedimentary rock layers show evidence of deposition in water flowing fast or slow.[2] Visitors to Mount St. Helens can see sedimentary rock layers hundreds of feet thick that were laid down in minutes during the 1980 eruption. Nobody has demonstrated rock layers that form slowly. For that matter, worms and clams obliterate within months any thin layering that forms where mud does slowly accumulate, like river deltas. In the letter, this man actually wrote, "The earth was filled with billions of fossilized dead things, and they were all very good!!"[3]

Since when is death good? If we must call "good" the torturous death by drowning and choking (asphyxiation) that fossilized vertebrate animals found locked in rock layers suffered, then how do we explain God's motive for making a covenant[4] with Noah, saying, "The waters shall never again become a flood to destroy all flesh" (Genesis 9:15)? Even God thought that destruction of all flesh was terrible enough for Him to reassure Noah's family that they should no longer fear its recurrence. Again, if all we need is lots of water, not lots of time, to form vast rock layers, then we can place those fossilized dead animals after the originally death-free, very good creation. And we can give the words of Genesis their normal meaning while affording the Holy Spirit more understanding than any scientist about what happened in the past.

Those set on defending the awkward idea that God's curse on the ground (Genesis 3:17) applied only to humans and not to animals, stars, or any other created thing face stiff challenges. For example, wouldn't this view force us to bend the plain meaning of Romans 8:20-21?

> For the creation was subjected to futility, not willingly, but because of Him who subjected it in hope; because the creation itself also will be delivered from the bondage of corruption [i.e., enslavement to passing away] into the glorious liberty of the children of God.

Instead, at every turn, the text best fits with other passages and with the world. A child's basic understanding of Genesis 3 corroborates a straightforward take on Romans 8:20. The gist of "and the waters prevailed exceedingly on the earth, and all the high hills under the whole heaven were covered" (Genesis 7:19) corroborates "the world that then existed perished, being flooded with water" (2 Peter 3:6). And both of those global Flood passages make sense of the fact of fossils stuck in the land of every continent.

I have become increasingly convinced that recent creation and the global Flood do the opposite of hindering the gospel. They reinforce the gospel by showing full confidence in the entire Bible, by offering a solid historical basis for every gospel concept, by teaching and believing an internally and externally consistent message, and by interpreting passages using consistent principles. My respect for Genesis started to grow. And when I reflected on what Genesis says about the goings-on in my own heart, it grew even more.

On the Origin of Our Ugly Insides

I have grown to love the process of comparing the world with the Word. Every time—sometimes sooner, sometimes later—I learn how some kind of scientific discovery confirms Genesis. And every time my confidence in Scripture grows just a little more. It may come from reading evolutionists' struggles to describe exquisite design without God. They marvel, for example, how nature alone optimized human eyesight to collect photons from the entire surface exposed behind the eye lens and how eyes can detect single photons without having to bathe the detector in liquid nitrogen like man-made cameras. It may be reading about connective tissues and blood vessels in dinosaur or other fossils that are supposed to be millions of years old. Evolutionists marvel at how such fragile materials could have lasted that long, but a recent global flood that formed these fossils thousands, not millions, of years ago resolves the tension.

But that's just me. Not everyone is into science. The Lord is so gracious that He tailors His communication to each person. And all of us desperately need to hear and believe the truth that He has revealed in Genesis about where we came from. We don't need to earn science degrees to access that truth. We don't need much to verify that history for ourselves either. We could take a short cut by applying a little bit of logic to our inner thought lives.

Scripture tells it like it is about my sins. It gives the unvarnished truth about how wretchedly selfish I can be—how short I fall of God's glory (Romans 3:23). The person who has no access to the latest scientific discoveries still has access to his own soul. All he needs to do is compare what's going on inside his heart to what the Bible says about human hearts. "The heart is deceitful above all things, and desperately wicked; who can know it?" (Jeremiah 17:9).

We fool ourselves when we ignore the fact of our lawbreaking; for example, when we undress someone in our minds. We just committed fornication (if we are unmarried) or adultery (if we are married) in our hearts! The Lord Jesus exposed these truths like no other. He said, "You have heard that it was said to those of old, 'You shall not commit adultery.' But I say to you that whoever looks at a woman to lust for her has already committed adultery with her in his heart" (Matthew 5:27-28). God intends this privacy as a blessing only between spouses, not for corrupted onlookers. Importantly, such evil thoughts violate that man or woman by demoting them from image bearers to objects.

Thus, a Genesis-based origin of men and women supports his or her honor in more than one way. In the beginning, the first husband and wife were very good—pure with one another and before God, knowing no shame. The injection of human rebellion into this purity as recorded in Genesis 3 explains the cause of the shame we experience today. We also infer from Genesis that the man or woman in question belongs to God, not to you or me. And because God made each of us, it follows that He cares enough to demand justice for each of us. So, unrepentant undressers must pay the proper penalty.

We may fool ourselves again if we blast past God's signposts in our frenzy to have our own way. For example, do any of us still follow Eve's tracks? "The serpent deceived Eve by his craftiness" (2 Corinthians 11:3). Yes, but she allowed herself to be deceived. "Let no one deceive himself" (1 Corinthians 3:18). "Let no one deceive you with empty words, for because of these things the wrath of God comes upon the sons of disobedience" (Ephesians 5:6). Do we, as men or women, give our hearts permission to deceive ourselves into thinking that right is wrong and wrong is right if it would achieve our selfish desires?

Think of Christian gals or guys who choose to date that non-Christian, persuading themselves that the unbeliever will still somehow fulfill their deep relational needs. That decision made in a moment can bend toward a lifetime of regret. But we fall prey to self-deception even in small matters—though they may not feel so small to God. We may, for example, persuade ourselves that it's not wrong to sneak snacks into the movie theater. And then we invent an excuse like "they charge too much for their snacks" in just the same way Adam invented excuses when he justified his rebellion with "the woman whom You gave to be with me, she gave me of the tree, and I ate" (Genesis 3:12).

God cursed Eve because she had done wrong—same as the serpent and Adam. Alongside her failure to trust in her Creator and take Him at His Word, she deceived herself into thinking that God was deceiving her by keeping her eyes from being opened and from being like God herself (Genesis 3:5). When we deceive ourselves about God's character, our self-importance, or any of a million other matters, we demonstrate the truth that Genesis reveals about the desperate wickedness of the human heart.

No other "religious" text speaks so plainly and so truthfully about the ugly inside that we each know about ourselves. Of course, men and women deny or suppress this truth—yet another evil tendency of the human heart. This suppression compels us to deliberately forget facts.

For example, if we really did just wish that the guy who cut us off on the highway would die, then we committed murder in our heart! That person is made in God's image, no matter how they have affronted us. And how soon we forget the times we pulled in front of other drivers the same way.

The Bible tells the truth about what we know *in* our hearts *about* our hearts. And since the God who inspired prophets and apostles to write the Bible speaks such accurate truth about such a deeply personal realm, we can have confidence that He speaks accurate truth about any realm—including the history recorded in Genesis. Having reached this point in my journey toward Genesis and toward the God who claims responsibility for it, I turned my attention to another aspect of my inner thought life—my mental processors. Do Genesis origins have anything to do with the way that we process our world?

Genesis and Meaning

What does it take for anyone to know anything? Well, an ability to reason helps. With this tool in our uniquely human minds, we can find certainty in knowledge by letting logic remove self-refuting ideas. Stick with me on this. It may take a few paragraphs, but in the end we will link Genesis with certainty of meaning.

A self-refuting statement says one thing but assumes the opposite. It fails because it holds opposites as both true when in fact only one can be true. Like a round square. Not possible. Once you pick up on self-refuting statements, they start turning up all over the place, and it becomes downright fun to know for sure why to absolutely reject them.

Have you ever heard these truth claims? "Truth is relative." On the surface it states that truth is not the same for different people who may see things in different ways. Of course people see things in different ways, but beneath the surface the very assertion that "truth is relative" assumes that it is absolutely (not relatively) true that truth is relative. In other words, the statement itself makes an absolute truth claim, even while declaring that there's no such thing. This one fails the basic sensibility test.

How about "you shouldn't judge others"? Except that the statement itself is a judgment. Or "you should tolerate all views." Friends and neighbors repeat this as though it encapsulates a sage insight, but it demands tolerating the view that you should not tolerate all views. It basically says that you should tolerate all views, including the view that you should not tolerate the view that "you should tolerate all views." It's impossible to live and thus easy to reject. We find more causal self-refuting statements, too, like "I'm speechless." Apparently not.

Now let's outline where this leads. The opposite of a self-refuting statement is an undeniable one. Consider "I exist." I undeniably exist, for I have to exist in order to try to deny my own existence. Thus, "I do not really exist" refutes itself. It tries to hold up two contradictory truths. On the one hand it outwardly asserts that I do not exist, but on the other hand it implicitly assumes that I do exist in such a way as to make the statement. Since "I do not exist" self-refutes, we should consider "I exist" as undeniable.

What about "God does not exist"? Of course, before looking beneath the skin of this increasingly common comment, all parties must first agree on a definition of God. I'm talking about the God of the Bible. He exists apart from the created universe and time. He called the universe into being, controls its every motion, sustains it every moment, and can interact with the universe from His position outside it.

What would result from a universe in which God does not exist? Well, you would have no universe. Our universe, by definition and description, does not perform miracles and thus could not make itself. It does not create stars (Disagree? Then simply name any star that we have witnessed being born), craft worlds, build trees or bees, you's or me's out of dust. It doesn't even make the dust. In other words, without a God to form the universe and its inhabitants, including the person who says "God does not exist," we have no adequate cause for the universe or its statement makers.

Thus, the surface assertion that there is no God requires the existence of a person to say it. And since people don't make themselves, their existence demands a God to have made them. In this case, it appears that the statement "God does not exist" does not refute itself directly, but it does lead to self-refuting situations. For example, if we can't exist without a Creator to perform the supernatural processes required to overcome natural processes, then it becomes quite safe to conclude that God does exist after all—logically, at least. It turns out that most who claim God does not exist do so for reasons of emotional trauma instead of cold logic.

What has all this to do with Genesis? Genesis helps explain why we can know anything for sure. It says that God—all-wise, all-powerful, and all-loving—created the universe out of nothing. This contrasts with a Big Bang universe that its proponents pretend made itself. They say we are stardust that turned into people. If so, then why do these people discover truths using laws of logic? We wouldn't know anything if we ran around thinking of round squares, relative truths, or uncreated creations.

But we *can* know things. We can know some things with certainty, like undeniable truths—round circles and created people. Yes, the Spirit of truth "will guide you into all truth" (John 16:13), but why doesn't the Spirit of God reveal that same truth to flowers, fish, or ferrets? Because God didn't give them the same capacity for reason. We apprehend the truths available by means of this aspect of the image of God in each of us. Big Bang universes wouldn't supply those aspects. An all-wise God would know the laws He made. An all-powerful God would have the ability to implant those laws into His creations. An all-loving God would desire to implant those laws into those whom He wished to know intimately. Thus, we know that we can know certain things since certain knowledge is required for us to have a relationship with Him. Genesis introduces the facts that He is interested in relationships and that our sin keeps us from that relationship.

This chapter has the title "Genesis and Truth." In it, we walked through a few of my discoveries about how Genesis origins underpin essentials of human life. High regard for the details in the Genesis account helps realign the foundation for the gospel of our salvation from the penalty, power, and eventually the presence of sin. Genesis truths confirm our heartfelt truths about our own morality. A Genesis-based creation by God informs our understanding of how we can understand. It also introduces us to His Spirit and to our special origins in His very image. What began as an intellectual longing for resolution to the problems of Genesis morphed into something much more personal.

If Genesis underpins the gospel, then it makes sense that God superintended the authorship of both Genesis in the Old Testament and the gospels in the New. By taking as true what Genesis teaches about my moral and intellectual capacities as reflections of His image inside me, I began to enjoy not only the satisfaction of elevated trust in His Word as true but to enjoy Himself as the truth.

The weight of evidence for Genesis as history began to bottom out the scale. It reached a point for me of being what lawyers call proven "beyond reasonable doubt." Yes, Lord, I trust what You say in Your Word! Once that happened, my relationship with Him blossomed. In my journey, the God who I knew as Savior began revealing Himself as the God of my creation and as the divine Author whom I could trust wholeheartedly. I was living in light of Genesis with the deepening fellowship between a sinner and his Savior.

Ever since then, I have wanted other Christians to know their Savior in this way. For me, believing in Genesis was not a side issue. It was not just academic. It led to a new world of awe for my Creator. It led to a wholesale rewiring of how I think about every aspect of life—how to view living things, rock layers, what goes on inside my own heart, how society should be structured, and what I think about Him.

We now have more reasons to trust Him than we did a century ago, and yet our cultures trust Him less than before. Should we prepare ourselves to embrace and defend God's whole Word in the face of a society of Bible deniers? I offer some reasons why we should do just that in the final chapter.

7
WHY GENESIS MATTERS MORE THAN EVER

"For if you believed Moses, you would believe Me; for he wrote about Me. But if you do not believe his writings, how will you believe My words?"
(John 5:46-47)

The Genesis Disconnect

Why do so few people take Genesis at face value? Well, we have pointed out the practical worldview or doctrinal aspect of that question. If evolution is true, then Genesis cannot be true since they make such opposite claims about the past. But what about a spiritual aspect to that question?

Sophisticated Spiritual Strategy

If there were a devil, would he not relish opportunities to dissuade humans—those miserable, weak images of the hated Creator—from believing Genesis? Simon Peter warned new Jewish believers in Christ to "be sober, be vigilant; because your adversary the devil walks about like a roaring lion, seeking whom he may devour" (1 Peter 5:8). What better way to devour spiritual lives than to persuade them that God did not in fact create them but nature did?

Thinking strategically, a devil who propagates the idea of death long before sin would effectively annul the basis for Christ's death for sinners. By erasing the foundations for our origins and purpose—which centers on knowing and enjoying God forever—devilish doctrines disconnect us from Genesis. To connive a way for folks to feel fully assured that experimental science backs these imagined doctrines must feel especially delectable to devils who live to deceive.

The Bible does teach that the devil's demons spread their ideas among rebellious men and women.

> Now the Spirit expressly says that in latter times some will depart from the faith, giving heed to deceiving spirits and doctrines of demons, speaking lies in hypocrisy, having their own conscience seared with a hot iron, forbidding to marry, and commanding to abstain from foods which God created to be received with thanksgiving by those who believe and know the truth. (1 Timothy 4:1-3)

We have evolved past those faded and rudimentary concepts like marriage, they argue. Traditional family structure comes from a disproven and laughable book called Genesis. Join our enlightened society, they say, which embraces all forms of true love, including any form of gender you'd like to invent for yourself! Instead, we need desperately to embrace the God of that Genesis we so easily disdain. He would heal us. He turned me right side up, for one.

So Uncool

In addition to a worldview and spiritual aspect, a personal pride aspect adds its influence. In the introduction to his book *The Potter's Promise*, Bible professor Leighton Flowers describes the goings-on of his inner man as he entered the painstaking process of reexamining what the Scriptures say about free will and predestination. Flowers wrote, "Our innate desire to be esteemed by others and seen as 'smarter' than we really are often overwhelms any potential for learning and profitable dialogue."[1] That same innate desire manifests in conversations over origins issues.

Whoever decides to take Genesis at face value faces a bevy of repercussions. You have to somehow defend creation over evolution, and evolution is backed by a world full of really smart scientists. You have to somehow defend recent creation over billions of years, and billions of years is (supposedly) backed by a world full of really old rocks. But worst of all, you have to endure the ridicule of a world full of folks who stand convinced that Genesis is a joke. Nobody wants to be laughed at.

In the introduction to their book *Chasing Cool*, which centers on marketing, authors Noah Kerner and Gene Pressman summarized this intense social pressure. They wrote, "Indeed, our society is *consumed* with the trappings of cool....All across the psychographic spectrum everyone wants it, even if they can't define what 'cool' actually is."[2] But we know what cool is not. It's not cool to stick with the supposedly outdated and unnecessary myths in Genesis.

Kerner and Pressman are right, and so is Flowers. Everyone wants cool, even—or perhaps especially—academicians. Everyone wants to be the smart one in the room. It's hard to feel accepted and appreciated when you take a position that so many others disdain. Especially when those others are your family, friends, or your boss. What's the solution?

Reorient your soul. It may seem hard, but in the end what are social pressures compared to the overwhelming sense of the fear of and relationship with God that we should embrace? "The fear of the LORD," not the fear of rejection, "is the beginning of wisdom" (Proverb 9:10). Find joy, strength, sense of purpose, and sense of significance and belonging in the Lord Jesus and nobody else. In short, "turn away my eyes from looking at worthless things, and revive me in Your way. Establish Your word to Your servant, who is devoted to fearing You" (Psalm 119:37-38). After all, this world with its evil works and its disdain for the Creator and His Word will not last. The Creator and His followers will.

Genesis and Everyday Life

Another reason why Genesis matters more than ever is because of the way that a full confidence in Genesis would elevate our everyday lives. We have generations growing up with no knowledge of their origins. They learn that they are merely hairless apes—the result of pointless, purposeless processes that produced people from particles. Re-educate that kind of mind with Genesis origins and then ask what aspect or angle in all of life would be left untouched?

Genesis creation says I came from God, not nature. I therefore owe my breaths to Him. Just like with Adam and Eve, He knows everything I do and even think. He made me for a purpose. I am now encouraged to find out what that is and how to fulfill it. How wonderful to learn that when I do just that, I discover that His yoke is easy, His burden light, and the joy of the Lord is my very strength (Matthew 11:30; Nehemiah 8:10)!

Genesis pinpoints God as the inventor of marriage. It's His idea, not that of any animal ancestors. Living in light of Genesis affirms that husbands and wives stand equally accountable before God for their individual actions. Genesis 3 says, "Because you have done this," not "Because your spouse has done this." God will therefore hold me personally accountable for my own sins.

Genesis teaches that Eve was not taken from Adam's foot as though she were any bit the lesser but from his side as his equal before God. How many failing marriages could this understanding heal where one gender believes they are superior to the other? Living in light of Genesis would lead wives to refrain from exploiting their husbands' inferior verbal skills just to win an argument. Nor would husbands any longer take advantage of their wives' emotional or physical differences.

Genesis indicates just the opposite. My wife is my equal before God. In fact, God crafted our strengths to cover—not exploit—one another's weaknesses in marriage. Husbands should thus respect and inquire of her choices, and wives should respect their husbands' input. The two begin to become one, as intended.

No observant American would deny the general tragedy of an increase in broken lives and broken homes in *every* cultural demographic. Each sad metric, whether one looks at likelihood to end up with a harmful addiction, lower lifetime salary earnings, or increased feelings of dissatisfaction and isolation, ties to broken homes. We will witness even more carnage from the wake of abandoned families.

God invented families to have husbands live with wives based on commitment to one another—a commitment that our Genesis-scoffing culture understands less and less. Social statistics confirm that following in the marriage pattern of Adam and Eve gives the very best chance for parents to maintain the stability of their marriage and for kids to grow to live healthy, successful, and fulfilled lives. Healthy families produce healthy people, and they create prosperous nations. Those effects could all become mere side benefits of Genesis-respecting, rightly oriented families that develop a healthy fear of, and passionate love for, the Lord as Creator and Savior.

Dwindling Christianity

Genesis Beliefs Predetermined

Christianity is losing respect in our Western cultures. With its foundation of Genesis relegated to mythology, belief in the Christ who knows Genesis as real history becomes a bit of a joke. But that's nothing new and does not surprise the Creator and Sustainer of the universe. Scripture gives examples of this and speaks of unbelievers' disdain for God and His Word.

The great Welsh 17th-century preacher and Bible commentator Matthew Henry summarized the Bible's view of those who refuse to believe, saying:

> Thus is the power of the word in many baffled by the power of prejudice: they do not believe, because they are resolved they will not: they conclude that no good thing can come out of Nazareth, John i. 46. and will not be persuaded to come and see. Thus do they prejudge the cause, Prov. xviii. 13. 'answering the matter before they hear it,' and it will prove folly and shame to them.[3]

Some will refuse to trust Christ's words no matter how reasonable they are and no matter how accurately they explain the real world, including evidence for recent creation or for the sinful workings of our inner hearts. And yet that does not describe every lost person. I was lost but now am found. I was blind to the importance and trustworthiness of Genesis, but now I see. I needed someone to explain to me the science that supports Scripture. Once I saw that, in part by reading the classic book *Scientific Creationism* by ICR's founder Dr. Henry Morris, the door opened for me to reconsider the historicity of the book of beginnings. I'm not the only one with such a turnaround.

Just because some will not hear reason does not mean all will refuse it. Most of the religious Jews denied the Christ, even though He performed undeniable miracles right in front of them. But some believed! Therefore, God directs His followers to trust and defend Genesis. He admonished through his apostle Peter, "But sanctify the Lord God in your hearts, and always be ready to give a defense to everyone who asks you a reason for the hope that is in you, with meekness and fear" (1 Peter 3:15). In context, Peter was preparing his readers to persevere through persecution. When Christ-followers get rounded up, tortured, and executed in the name of the Lord they follow, they should ready themselves to defend why they cling to the Lord and His Word even unto death. Why? Because some will listen to reason, turn from wickedness, and ask the Lord to save. Some will repent and enjoy the Lord forever.

Christianity may not be on the rise just now. Respect for Genesis as the foundation for every New Testament doctrine and the whole of God's Word is reaching a new low. Therefore, the people of God need more than ever to return to Genesis with the same fervor the psalmist had when he clung to the Scriptures, saying, "The entirety of Your word is truth, and every one of Your righteous judgments endures forever" (Psalm 119:160).

What I Learned in Sunday School

One even more important issue I wish to emphasize in this section, other than Christianity dwindling in our broader cultures, is Christianity dwindling within our own churches! Too many Christians take the atheists' word as being more authoritative on matters of the past than the Word of God. Christians are losing ground with each generation that hears Genesis in terms of "Bible stories" instead of as history. Only a few young people stay in church after college anymore. Why should they stay if they have no idea how much better the Bible's origins in Genesis explain the real world than do atheistic speculations that sprinkle evolutionary textbooks? Why should anyone expect our children to cling to a crumbling, man-made religion?

Of course, Christ, not man, made the real Church—that collection of Christ-followers that began in Acts during the first century and has never died. But that's the point. Our young people are told that science proves millions of years and evolution, and that superstitious words of ignorant men invented Genesis. And older Christians do nothing. We just let them slip away, like watching our precious children zip from shore along a riptide to the horizon.

Sunday school teachers and, much more importantly, mothers and fathers need more desperately perhaps than ever to obey verses like 1 Peter 3:15 quoted above or 2 Corinthians 10:5, which says in part that we must "[cast] down arguments and every high thing that exalts itself against the knowledge of God." What particular arguments does this culture bring against the knowledge of God? Of course, our world is filled with such arguments, but I can think of none more popular than arguments about how science supposedly disproves Genesis. Talk about exalting itself against the knowledge of God—the removal of Genesis deletes the most basic characteristic of God. *He* made me, not nature. How do I know?

Well, the Bible does tell me so, but so do plain observations. What natural process ever made a new creature? Creatures always come from preexisting creatures. Life begets life.[4] But our public school textbooks all teach that life arose from non-life, and we let this nonsense slip into our children's minds unquestioned! What a shame. These same books, written by naturalists, also teach that nature can morph one creature kind into a completely different kind. More nonsense. This morphing only occurs in naturalists' imaginations. In the real world, creatures always arise from like creatures, just as Genesis says.

Parents and pastors would do well to take those under their care to visit a highway road cut or other nearby rock layer exposure. There by the roadside, why not talk about easy-to-see clues of rapid catastrophe? These point to Noah's Flood in contradiction to the slow and everyday processes that naturalists imagine.

Fossils form one such clue. Those don't happen today—at least not the way it must have happened in the past where mostly sea creatures got trapped in mud that turned to stone faster than the seashells or bones could rot. These clues call for an unimaginably terrific catastrophe that Genesis 6–9 explains.

Or what about those flat contacts between layers? Natural erosion processes carve gullies and rills into Earth's upper surface, but naturalists insist that these surfaces remained perfectly flat along hundreds of miles for countless years until the next layer gently landed atop it!

Such a scene happens nowhere in today's normal world. Abnormally fast-moving masses of mud explain it. The Genesis Flood accounts for flat-topped sedimentary layers that extend for hundreds of miles. And if the Genesis creation and Flood really happened, then the God-become-man who quoted Genesis really spoke the truth. I only *wish* I had learned *that* in Sunday school.

The Limp Church

Christian after Christian, including me for a while, limps through our spiritual lives when we were meant to walk in victorious confidence. The church in the West suffers from several systemic diseases, but the one I am trying to treat with this book is that lack of confidence in God as the One who wrote and meant every word in Scripture. This disease impacts core activities of the Christian life, starting with evangelism.

Genesis and Evangelism

How effectively can a Christian believer share the Bible's good news if he or she feels unsure about which verses God meant and which ones God either got wrong or somehow meant something other than what He inspired His prophets or apostles to write? It's like a salesman trying to sell you a product he doesn't believe in.

My friend Don did that. His experience illustrates this disease of the church. He lasted as a door-to-door encyclopedia salesperson for only about a week. He *almost* sold one book set to an extremely poor man who lived in a shanty. Don knew that man should spend what little resources he had on food and shelter instead of luxuries like encyclopedias. As the man began to agree to the book set, Don dissuaded him, saying, "I really don't think you need this." Don sold zero encyclopedias and found work elsewhere.

How many times have words to that effect scrolled across the screen of Christian minds when they think of sharing the gospel? "I really don't think you want all this Bible. Some of it is untrue." All because the atheistic religious tenets of naturalism, masquerading as science, has cowed Christians into distrusting Scripture.

How can we preach the gospel's message of sin if we erase the origin of sin found in Genesis? How can we explain the gospel's message of accountability to a Creator if nature, not God, created everything after all? How can we answer questions about Christ's payment of our death penalty when we have smuggled into our minds through the assignment of vast ages to fossils the atheistic doctrine that death came not from sin but from some universal principle of evolution?

There is a better way. A way that jives with both Scripture and science. It takes time to sift through the raw results of data versus their evolutionary interpretations. It takes time to sort through the Scriptures and let what God says about the world reprogram our minds. But the end result is a whole and entire, resolved and complete, defensible and satisfying worldview.

We should preach, explain, and defend the gospel. To do this well, we must each one take full confidence in the whole Bible—"*all* things that I have commanded you" (Matthew 28:20)—as the Word of God. It does not merely contain His words like a field of wheat mixed with weeds, but "*all* Scripture is given by inspiration of God and is profitable..." (2 Timothy 3:16).

Now I can approach any person in any walk of life with confident assurance that God made us for fellowship with Him even as Adam walked with God in the cool of the day in the beginning. Although our sins separate us from Him, He handled those trespasses so that we could access the glorious garden of the future and the infinitely more glorious God who crafted us to know and enjoy Him forever.

Genesis and Prayer

Prayer comprises another core activity of the Christian life. How well can we pray without that Genesis foundation?

Prayer is all about talking with God. But who is He? I would presume of God less and honor Him more were I convinced that He is the One who made me. Removal of Genesis, *a la* atheistic or theistic evolution, equates to removal of the most basic aspect of God. Since He made me, I owe Him everything.

Removal of parts of Genesis, *a la* progressive creation as discussed above, leaves God's core identity as the Creator intact, but it lessens some of His other aspects. For example, who is God if He is not truth? I need—we each need—to be able to trust God's truthfulness. But what if He didn't tell the truth when He says that the reason we all take one day per workweek to honor Him is because "in six days the LORD made the heavens and the earth, the sea, and all that is in them, and rested the seventh day. Therefore the LORD blessed the Sabbath day and hallowed it" (Exodus 20:11)? Either He is truthful and trustworthy or not at all. Scripture leaves no room to compromise God's character by making certain words mean what we want them to mean instead of letting all of His words mean what He intended based purely on what He wrote.

The varied Genesis compromise views noted in chapter 2 count Him as Creator, but if the prophets and apostles took Genesis at face value when "we now know" that Genesis should no longer be taken straightforwardly, then did this Creator mislead those prophets and apostles whom He used to write the Bible? Did this Creator mislead me by allowing incorrect history to creep into what is supposed to be His correct Word? Now, what kinds of prayers would I direct to a God who is either not capable or not willing to tell me the exact truth about where I came from? They might sound like this: "Lord, please make whatever small contribution that Your slight capabilities might offer toward this problem." Yuck. Talk about lukewarm.

God's nature and identity flow from His very Word. Wherever we struggle with what He says, we struggle with the nature of God. A born-again believer who walks in full assurance of God's total Word prays to a God who made everything, yes. But she also prays to a God who made everything how and when He said He did—that is, by His spoken command thousands, not millions, of years ago. That is the God of the Bible. And what a joy to pray to the One who holds power to make and sustain the universe and yet cares enough to assure little old me of my origins from Him, the origin of sin, and His glorious salvation through Jesus' death and resurrection.

Joys from Genesis

God, People, and Trees

One of the joys that accompanies acceptance of Genesis as a true account of our origins is the never-ending series of discoveries it affords. Its precisely written words point to God's inexplicable love for me. Take trees, for example.

Not as much trees in general, but I mean specific trees like those in the Garden of Eden. Before the Flood destroyed all of Eden along with the whole world, God walked and talked with His newly created people in a garden. It was life-filled and tall-treed, perhaps like what J. R. R. Tolkien was going for in his elven cities.

> The LORD God planted a garden eastward in Eden, and there He put the man whom He had formed. And out of the ground the LORD God made every tree grow that is pleasant to the sight and good for food. The tree of life was also in the midst of the garden, and the tree of the knowledge of good and evil. (Genesis 2:8-9)

People walked and talked with the living God in the garden. God spoke His first recorded words with Adam while they walked among the garden's trees.

God had Noah rescue food-bearing trees and other plants from the Flood judgment. "And you shall take for yourself of all food that is eaten, and you shall gather it to yourself; and it shall be food for you and for them" (Genesis 6:21). That has made countless dinner tables taste delectable. It also made it possible for Him to appoint a particular tree that seems to serve as a symbol of His presence.

About 2,000 years after creation, God spoke to Abram, promising that his descendants would fill and live forever in a land of God's promise. "Then Abram moved his tent, and went and dwelt by the terebinth trees of Mamre, which are in Hebron, and built an altar there to the LORD" (Genesis 13:18). Terebinth trees are almond trees. So, in both the garden and in Mamre, God and man met and spoke amidst trees.

God met mankind in the tabernacle some 430 years after Abraham. It was a mobile meeting place into which God's holy and fiery presence descended from heaven. God would talk with the high priest in that golden center of the tabernacle—and later, the temple in Jerusalem—called the Holy of Holies. His people could hear from and know Him.

But—this time quite unlike Tolkien's fantasy world of walking trees—the Israelites could not carry around a planted garden. So, God gave specific instructions for the artwork on golden vessels meant to adorn the Holy of Holies, saying, "Three bowls shall be made like almond blossoms on one branch, with an ornamental knob and a flower, and three bowls made like almond blossoms on the other branch, with an ornamental knob and a flower—and so for the six branches that come out of the lampstand" (Exodus 25:33). Here again near the likeness of trees, a holy God fellowshipped with wicked people.

Later, "the word of the LORD came to me, saying, 'Jeremiah, what do you see?' And I said, 'I see a branch of an almond tree.' Then the Lord said to me, 'You have seen well, for I am ready to perform My word'" (Jeremiah 1:11-12). His word spoke of judging His rebellious nation. He did perform His word when Nebuchadnezzar destroyed Jerusalem in 587 BC.

Then the Lord Jesus—the perfect Creator made man—gave up His own life in order to bring you and me back to God. "Now in the place where He was crucified there was a garden, and in the garden a new tomb in which no one had yet been laid" (John 19:41). God's own Son paid the penalty that our sins earned. The Son of God was buried amidst a garden, like a planted human seed. And a tree of new and transformed lives began to shoot up when the Lord rose from the grave. Its branches wind through history from that day until this. Each sinner who repents of sin and trusts in Jesus Christ as Savior and Lord forms a new blossom on God's growing tree. What joy to know that I was part of God's plan from the beginning! How wonderful to know that the Lord had me in mind when He planted references to trees in His Word, beginning in Genesis.

Finally, God will eventually make new heavens and a resurrected earth into which He will place a new golden city with a grand garden for the redeemed. What will happen there? God will once again walk and talk with men and women—those whom He bought back to Himself through His Son's sacrifice. Forever afterward we will enjoy a place, adorned with trees, of intimate fellowship with our Creator and Redeemer. "In the middle of its street, and on either side of the river, was the tree of life, which bore twelve fruits, each tree yielding its fruit every month. The leaves of the tree were for the healing of the nations" (Revelation 22:2). A return to the garden. But this time, it's a forever good place with forever fellowship.

One of the joys of living in light of Genesis is the discovery of ways—and pictures of those ways—that the Savior reconnects with rebels like me.

The Ultimate Joy

I've been trying to urge my dear reader toward a return to that Garden of Eden. Science confirms Genesis history, and universal human truths confirm Genesis tenets, but those considerations are largely academic. A return to Genesis—to that originally very good place—is to return to the Creator Himself. The ultimate joy in life, of life, and beyond this brief life is to know Him forever. To know Him is to take Him at His word. As the psalmist wrote:

> I rejoice at Your word
> As one who finds great treasure.
> I hate and abhor lying,
> But I love Your law.
> Seven times a day I praise You,
> Because of Your righteous judgments.
> Great peace have those who love Your law,
> And nothing causes them to stumble.

> Lord, I hope for Your salvation,
> And I do Your commandments.
> My soul keeps Your testimonies,
> And I love them exceedingly. (Psalm 119:162-167)

Rejoicing at His Word, hating falsehood, loving His law, praising His goodness, finding peace and stability in His law, hoping for a sure salvation, doing what He says to do, and keeping and loving things that He says all come from knowing God personally—not merely knowing about Him. One can even give intellectual assent to Genesis as history but still not have a relationship with God. Clearly, the psalmist was enthralled with the Lord Himself, and all these aspects flowed from that relationship.

One way to express the ultimate joy that I'm trying to treat here is to contrast His Word with the world's perspective on origins. For example, consider how the world might render Psalm 102:25: "Of old ~~You~~ Nature laid the foundation of the earth, and the heavens are the work of ~~Your~~ Nature's hands." Either God got this wrong, or the world has it upside down. Living in light of Genesis means we don't have to worry any more about God getting it wrong. The psalmist nailed it when he said, "The heavens are the work of Your hands!" It means we can trust His every word. Since He knows best and knows everything, we take great joy in embracing His truths!

Why should each of us live in light of Genesis and with the God who wrote it for us? Only with Him will we find our purpose for being. Only there, walking and talking with God our Maker and Redeemer, can the broken pieces of our selfish lives get reassembled into honorable and whole humans. Only there among His trees that bear the fruits of joy, peace, inner health, and everlasting wholeness and where we trust in Him alone do our questions find answers, our problems find resolution, and our anxieties find rest.

But it's not really about us. It's about Him. The ultimate joy comes from fellowship between two people—between the sinful man or woman who reunites with this holy God who has pursued us all our lives. Then, everything has new meaning. Evening skies become God's canvas upon which He paints a stunning and fiery mural *for us to enjoy in that moment*—not so we can merely say, "Wow," as we once did when we knew no appropriate recipient for our stillborn praise, but so we can say, "Wow, fantastic, God!"

Fathers and mothers who live in light of Genesis—who return to God through the Lord Jesus Christ—can embrace their children not merely as winners of the next round in a cruel Darwinian lottery but as dearly beloved little ones made in the image of God with all the potential to know and enjoy Him forever.

I say, why not? Why not trust Him with every concern, "for He cares for you" (1 Peter 5:7)? Why not live in light of Genesis by humbling ourselves? "'For all those things My hand has made, and all those things exist,' says the Lord. 'But on this one will I look: on him who is poor and of a contrite spirit, and who trembles at My word'" (Isaiah 66:2).

Genesis is the foundation of God's Word. When we believe what He says, then He looks on us as a good Father and smiles. Nothing beats that.

ENDNOTES

Chapter 1: What Genesis Says
1. Luther, M. 1959. *What Martin Luther Says: A Practical In-Home Anthology for the Active Christian*. E. M. Plass, ed. St. Louis, MO: Concordia Publishing House, 93.
2. Anonymous to Brian Thomas, September 24, 2016.

Chapter 2: Genesis and Evolution
1. Zindler, F. 1996. Atheism vs. Christianity (debate). Video, Zondervan.
2. Maisey, J. 1996. *Discovering Fossil Fishes*. New York, NY: Henry Holt & Co., 217.
3. They include John Walton, William Craig, Peter Enns, and John Collins.
4. Whoever might balk at this needs only logic to recall how even vile sinners remain able to correctly solve math problems.
5. Drake, M. 2017. *The Misted World of Genesis 1*. Auckland, New Zealand. Wycliffe Scholastic. Kindle Edition, 113.
6. "The clarity with which Genesis insists that God is eternal and that the cosmos is not, that God is distinct from the cosmos, that things within the cosmos are differentiated and not transmutable, and that the cosmos is made while God is not, makes it as impossible for paganism to be the interpretive framework for the Genesis account of creation as it is impossible for paganism to conceive of creation." Ibid, 70.
7. Guliuzza, R. J. 2021. Walton's Lost World Obscures Biblical Clarity. *Acts & Facts*. 50 (7): 6.
8. Drake, 133.
9. Ibid, 131.
10. Sibley, A. 2018. Deep time in 18th-century France—part 1: a developing belief. *Journal of Creation*. 32 (3): 87.
11. Ibid.
12. Hume, D. 1779. *Dialogues Concerning Natural Religion*, 2nd ed. R. Popkin, ed. London: Hackett Publishing Company, Inc., 141.
13. Darwin, E. 1794. *Zoonomia: or the Laws of Organic Life*, vol. 1. London: Printed for J. Johnson in St. Paul's Church-Yard, 509.
14. Lyell, C. 1881. *Life, Letters, and Journals of Sir Charles Lyell, Bart*. vol. 1. K. M. Lyell, ed. London: John Murray, 271.
15. Ibid.
16. Oard, M. 2019. *The Deep Time Deception*. Powder Springs, GA: Creation Book Publishers, 485.
17. Darwin, C. 1873. Darwin Correspondence Project, Letter no. 9105. University of Cambridge. Posted on Darwinproject.ac.uk.
18. Clarey, T. 2020. *Carved in Stone: Geological Evidence of the Worldwide Flood*. Dallas, TX: Institute for Creation Research.

Chapter 3: Science That Supports Genesis

1. Lyell, C. 1881. *Life, Letters, and Journals of Sir Charles Lyell, Bart.* vol. 1. K. M. Lyell, ed. London: John Murray, 271.
2. Genetic divergence of man from chimp has aided human fertility but could have made us more prone to cancer, Cornell study finds. *Cornell Chronicle*. Posted on news.cornell.edu May 16, 2005, accessed May 20, 2019.
3. Tomkins, J. P. 2018. Separate Studies Converge on Human-Chimp DNA Dissimilarity. *Acts & Facts*. 47 (11): 9.
4. Steinmann, A. E. 2011. *From Abraham to Paul: A Biblical Chronology*. St Louis, MO: Concordia Publishing House, 88.
5. Johnson, J. J. S. 2018. Viking Bones Contradict Carbon-14 Assumptions. *Acts & Facts*. 47 (5): 21.
6. Wood, B. G. 1990. Did the Israelites Conquer Jericho? A New Look at the Archaeological Evidence. *Biblical Archaeology Review*. March/April 1990: 44–58.
7. Tenev, T. G., J. Baumgardner, and M. F. Horstemeyer. 2018. A Solution for the Distant Starlight Problem Using Creation Time Coordinates. *Proceedings of the International Conference on Creationism*. 8: 82–94.
8. Faulkner, D. 2014. Response to: "Critique: Faulkner's Miraculous Translation of Light Model Would Leave Evidence." *Answers Research Journal*. 7: 461.
9. Objectors point to the fact that any two humans have far too many DNA differences for 100 mutations per 20 years at 6,000 years to have formed. However, this assumes that all difference arose from mutations. Instead, God would have embedded the majority of our differences, called common variants, into Adam and Eve's original DNA. This afforded all their descendants plenty of helpful variations. Only the rare variants have arisen since creation, and they tend not to help so much.
10. Drake, M. Human beings on brink of achieving IMMORTALITY by year 2050, expert reveals. *Express*. Express.co.uk, February 19, 2018.
11. Genesis 2:17.
12. Genesis 10:32.
13. Tomkins, J. 2015. Empirical genetic clocks give biblical timelines. *Journal of Creation*. 29 (2): 3–5, referencing Jeanson, N. 2014. New Genetic-Clock Research Challenges Millions of Years. *Acts & Facts*. 43 (4): 5–8.
14. Hodge, M. J. S. 1992. Natural Selection: Historical Perspectives. *Keywords in Evolutionary Biology*. E. F. Keller and E. A. Lloyd, eds. Cambridge, MA: Harvard University Press, 263.
15. Talbot, S. L. Can Darwinian Evolutionary Theory Be Taken Seriously? The Nature Insitute. Posted on natureinstitute.org May 17, 2016, accessed September 14, 2018.
16. Zimova, M. et al. 2014. Snowshoe hares display limited phenotypic plasticity to mismatch in seasonal camouflage. *Proceedings of the Royal Society B*. 281 (1782): 4.
17. Barrett, R. D. H. et al. 2010. Rapid evolution of cold tolerance in stickleback.

Proceedings of the Royal Society B. 278 (1703): 233.
18. Atlantic snails are increasing dramatically in size, Queen's researcher discovers. Queen's University press release. Posted on phys.org March 24, 2009.
19. Singer, E. The Legendary Biologists Who Clocked Evolution's Astonishing Speed. *Wired.* Posted on wired.com October 15, 2016. Accessed August 16, 2023.
20. The published decay rate for bone collagen is 183kJ/mol. However, my recent experiments indicate an even higher decay rate that would spell an even shorter shelf life.
21. Thomas, B. and S. Taylor. 2019. Proteomes of the past: the pursuit of proteins in paleontology. *Expert Review of Proteomics.* 16 (11–12): 881–895.
22. Thomas, B. 2019. *Ancient and Fossil Bone Collagen Remnants.* Dallas, TX: Institute for Creation Research.
23. Thomas, B. and B. Enyart. List of Biomaterial Fossil Papers (maintained). Posted on docs.google.com.
24. Gould, S. J. 1994. In the Mind of the Beholder. *Natural History.* 103 (2): 14.
25. Dirks, P. et al. 2017. The age of *Homo naledi* and associated sediments in the Rising Star Cave, South Africa. *eLife.* 6:e24231.
26. New Research Shows Early Ancestor May Have Coincided with Modern Humans. LSU Media Center. Posted on lsu.edu May 9, 2017.

Chapter 4: Culture Without Genesis

1. Spurgeon, C. H. 1889. *Second Series of Lectures to My Students: Addresses Delivered to the Students of the Pastors' College, Metropolitan Tabernacle.* New York, NY: Robert Carter and Brothers, 86–87.
2. *Evolution vs. God Uncensored* – Expanded and Updated. Living Waters. Posted on YouTube April 24, 2013.
3. Editorial: A scientist's case against God. *The Independent (London)*, 17. April 20, 1992.
4. How the retina works: Like a multi-layered jigsaw puzzle of receptive fields. Salk Institute for Biological Studies press release. Posted on salk.edu April 7, 2009.
5. Public's Views on Human Evolution. Pew Research Center. Posted on pewresearch.org December 30, 2013, accessed January 13, 2014.
6. Newport, F. In U.S., 46% Hold Creationist View of Human Origins. Gallup. Posted on gallup.com June 1, 2012.
7. Swift, A. In U.S., Belief in Creationist View of Humans at New Low. Gallup. Posted on gallup.com May 22, 2017, accessed July 24, 2017.
8. Lewis, C. S. 1991. *The Inspirational Writings of C.S. Lewis: Surprised by Joy.* New York, NY: Inspiration Press, 97.
9. Walvoord, J. 1969. *Jesus Christ Our Lord.* Chicago, IL: Moody Publishers, 13.
10. Luke 3:38.
11. Zimmerman, A. 2009. Marxism, law and evolution: Marxist law in both theory and practice. *Journal of Creation.* 23 (3): 90–97.
12. Engels, F. 1950. *Selected Works, 3 vols.* New York, NY: International Publishers, 153.

13. Taylor, A. 1989. The Significance of Darwinian Theory for Marx and Engels. *Philosophy of the Social Sciences.* 19 (4): 409.
14. Johnson, P. E. 2000. *The Wedge of Truth: Splitting the Foundations of Naturalism.* Downers Grove, IL: Intervarsity Press, 107.
15. Bergman, J. 1999. Darwinism and the Nazi race Holocaust. *Journal of Creation.* 13 (2): 101.
16. Weikert, R. 2004. *From Darwin to Hitler: Evolutionary Ethics, Eugenics, and Racism in Germany.* New York, NY: Palgrave Macmillan.
17. Haeckel also earned some infamy for ignoring differences and exaggerating similar features in a series of embryo illustrations intended to prop up the idea that mankind descended from animals. Modern photographs published in *Science* in the 1990s showed the same animal embryos in the same developmental stages in which Haeckel sketched them. The photos revealed that Haeckel totally ignored huge differences between the animals—for example, omitting an entire yolk sac in one drawing. They also revealed that Haeckel selected from different developmental stages, apparently to suit the desire to show similarities that might back up the supposed relatedness of all the quite different forms. Despite this, modern public school textbooks stunningly still print versions of Haeckel's embryo drawings as evidence for evolution.
18. Guliuzza, R. 2015. Major Evolutionary Blunders: The Eugenics Disaster. *Acts & Facts.* 44 (11): 10–12.
19. Dickens, C. 1958. *The autobiography of Charles Darwin 1809–1882: With the original omissions restored. Edited and with appendix and notes by his grand-daughter Nora Barlow.* N. Barlow, ed. London: Collins Clear-Type Press, 94.
20. Thomas, B. 2015. Evolution's Top Example Topples. *Acts & Facts.* 44 (10): 16.
21. Wood, B. G. The Discovery of the Sin Cities of Sodom and Gomorrah. Associates for Biblical Research. Posted on biblearchaeology.org April 16, 2008, accessed September 13, 2019.
22. Ganna, A. et al. 2019. Large-scale GWAS reveals insights into the genetic architecture of same-sex sexual behavior. *Science.* 365 (6456): 935–943.
23. Justice, T. CNN Claims There's No Way To Tell If A Baby Is A Boy Or A Girl. *The Federalist.* Posted on thefederalist.com March 31, 2021, accessed April 10, 2021.
24. Cole, D. South Dakota's governor issues executive orders banning transgender athletes from women's sports. *CNN.* Updated on cnn.com March 31, 2021, accessed August 6, 2021.
25. Relativism is the self-refuting philosophy that truth is merely a choice.
26. Aviles, G. Always to ax female symbol from sanitary products packages in nod to trans users. *NBC News.* Posted on nbcnews.com October 21, 2019, accessed August 6, 2021.
27. Guliuzza, R. 2021. Refusing to Live by Lies. *Acts & Facts.* 50 (5): 5–6.
28. For example, the Lord Jesus submitted to the Father unto death, saying, "Not My

will, but Yours, be done" (Luke 22:42).
29. Jones, R. K., E. Witwer, and J. Jerman. Abortion Incidence and Service Availability in the United States, 2017. Guttmacher Institute. Posted on guttmacher.org September 2019, accessed August 6, 2021.
30. O'Bannon, R. 62,502,904 Babies Have Been Killed in Abortions Since Roe v. Wade in 1973. *LifeNews*. Posted on lifenews.org January 18, 2021, accessed August 6, 2021.
31. Genesis 3:8.
32. Chirri, M. J. How Islam Views the Universal Creation. Ahlul Bayt Digital Islamic Library Project. Posted on al-islam.org, accessed September 11, 2019.
33. Protocol Additional to the Geneva Conventions of 12 August 1949, and relating to the Protection of Victims of International Armed Conflicts (Protocol I), 8 June 1977. In *Official Records of the Diplomatic Conference on the Reaffirmation and Development of International Humanitarian Law applicable in Armed Conflicts, Bern, Federal Department of Foreign Affairs*, 1978. International Committee of the Red Cross. Posted on ihl-databases.icrc.org, accessed September 11, 2019.
34. Milman, O. Lawyers argue Happy the elephant should have the same rights as humans. *The Guardian*. Posted on theguardian.com October 22, 2019, accessed October 22, 2019.
35. Frequently Asked Questions. Nonhuman Rights Project. Posted on nonhumanrights.org, accessed October 22, 2019.
36. Temple, J. Critics blast a proposal to curb climate change by halting population growth. *MIT Technology Review*. Posted on technologyreview.com November 5, 2019.
37. See the references in Hebert, J. 2019. *The Climate Change Conflict*. Dallas, TX: Institute for Creation Research.
38. Thomas, B. Does Earth Balance Carbon Dioxide Levels Automatically? *Creation Science Update*. Posted on icr.org January 12, 2009, accessed November 9, 2019.
39. Seuss, Dr. 1954. *Horton Hears a Who!* New York, NY: Random House.
40. Wilberforce, W. Speech to the House of Commons 1789. *The Abolition Project*. E2BN, 2007.

Chapter 5: Genesis and the Bible

1. Date estimated to be AD 30 from Lanser, R. What was the "Fifteenth Year of Tiberius"? Associates for Biblical Research. Posted on biblearchaeology.org May 15, 2019, accessed October 26, 2019.
2. Morris, H. M. 2012. Annotation for Mark 10:6. In *Henry Morris Study Bible*. Green Forest, AR: Master Books, 1476.

Chapter 6: Genesis and Truth

1. Lloyd-Jones, M. The Authority of the Bible. Sermon #5099. MLJ Trust. Posted on MLJTrust.org, accessed October 16, 2018.

2. Clarey, T. 2019. European Stratigraphy Supports a Global Flood. *Acts & Facts*. 48 (12): 10–12.
3. Anonymous to Brian Thomas, September 24, 2016.
4. A covenant is a solemn and authorized promise.

Chapter 7: Why Genesis Matters More Than Ever
1. Flowers, L. 2017. *The Potter's Promise: A Biblical Defense of Traditional Soteriology*. San Antonio, TX: Trinity Academic Press, 2.
2. Kerner, N., and G. Pressman. 2007. *Chasing Cool: Standing out in Today's Cluttered Marketplace*. New York, NY: Atria, xii. Emphasis in original.
3. Henry, M. 1853. *The Complete Works of Matthew Henry*, vol. 2. London: A. Fullarton and Co., 433.
4. This summarizes one of only two biological laws, called the law of biogenesis. Sadly, textbooks ignore it completely since it contradicts textbook discussions of abiogenesis—the never observed and forever imaginary formation of life from non-life. The second law of biology notes that like kinds beget like kinds—another direct refutation of evolution exactly noted 10 times in Genesis 1.

IMAGE CREDITS

Bigstock photo: 178

Thomas Cole: cover

ICR: 65, 75

NASA: 55

NPS: 70, 74

A. D. Riddle: 50, 51

Brian Thomas: 26, 68, 143, 145

ABOUT THE AUTHOR

Brian Thomas received a master's in biotechnology in 1999 from Stephen F. Austin State University, Nacogdoches, Texas, and a Ph.D. in paleobiochemistry in 2019 from the University of Liverpool. He taught junior high and high school at Christian schools in Texas as well as biology, chemistry, and anatomy as an adjunct and assistant professor at Dallas-area universities. In 2008 Dr. Thomas joined the Institute for Creation Research as a science writer and editor, contributing news and magazine articles, speaking on creation issues, and researching original tissue fossils. He was appointed as Research Associate in 2019 and Research Scientist in 2021. He is the author of *Dinosaurs and the Bible*; co-author of *Parks Across America: Viewing God's Wonders Through a Creationist Lens*, *Fascinating Creatures: Evidence of Christ's Handiwork*, *Human Origins*, and *Dinosaurs: Exploring Real-Life Dragons of History*; and a contributor to *Guide to Creation Basics*, *Creation Basics & Beyond: An In-Depth Look at Science, Origins, and Evolution*, *Guide to Dinosaurs*, *Guide to the Human Body*, *Guide to the Universe*, and *Dinosaurs: God's Mysterious Creatures*. His dissertation, *Ancient and Fossil Bone Collagen Remnants*, is available in book form.